THE AI-ERA ENGINEER

A Field Guide for the Next Generation

Kyle Benesh, PE

The AI-Era Engineer: A Field Guide for the Next Generation

Published by Benesh Engineering.

Paperback ISBN: 979-8-9960375-0-6
Hardcover ISBN: 979-8-9960375-2-0
Ebook ISBN: 979-8-9960375-1-3

theaieraengineer.com

First Edition

For the mentors — the ones who made time for the "why."

CONTENTS

PROLOGUE: THE PEN CLIP IN QATAR

The pin was bent.

We were standing in a mechanical room in the middle of the Qatari desert, five years into a project that had almost broken everyone involved. Multiple project managers had come and gone. The schedule had stopped and started so many times I'd lost count. We'd traveled 60 hours round trip just to be here for this commissioning window, and now a bent pin inside a VFD card slot was threatening to send us all home with nothing to show for it.

The mood was immediate and heavy. Everyone knew what a bent pin meant. You can't just order a replacement PLC card slot and have it show up in the Qatari desert by Tuesday. We were looking at weeks of delay, minimum. Another 60-hour trip back. More schedule slippage on a project that had already eaten years of people's lives. The VFD wouldn't run without that software card seated properly, and the card wouldn't seat with a mangled pin.

I looked at the damaged slot. Looked at my tool bag. Looked at the team.

"Fuck it. If it's already broken, I can't make it brokener."

I pulled the card out and went to work. It was like performing surgery in a mechanical room — tiny pin, tight clearance, hundred-degree heat, and the knowledge that if this didn't work we were done for the day, possibly for the month. I went through my tools one by one, looking for anything with the right geometry to reach the pin and bend it back to an angle where the card would seat.

About twenty-five minutes in, I found it. A pen clip. Just a standard pen clip that I'd pulled off and stuck in my bag at some point. It had exactly the right curve, the right thickness, and the right amount of flex to slide into the slot, hook the pin, and coax it back into position.

I slipped the software card back in. It seated. Fired up the VFD.

Back in business.

Later that day, we ran the world's largest vertical wind tunnel for the first time. I was one of the first five people to fly in it.

Let me say that again, because I want you to sit with it: I spent the morning performing microsurgery on a VFD with a pen clip in the Qatari desert, and by the afternoon I was flying through the air inside a machine I'd helped bring to life. That's engineering. Not the calculations. Not the software. The whole thing — the five years of persistence, the field troubleshooting, the improvisation, the team, and the moment where it all comes together and humans fly.

• • •

Here's the part that matters for you, the person reading this book.

I wasn't supposed to be there. There's no career path that says "mechanical engineering degree → wind tunnel commissioning engineer." That job didn't exist when I was in school. I made it exist by accident, through a chain of decisions that didn't look connected at the time.

I was a building designer in New York City. That's where I learned mechanical systems — boilers, chillers, ductwork, piping, controls. I worked at one of the oldest engineering firms in the city, learned from a mentor who'd designed submarines before he designed buildings, and built the foundation of everything I know about how physical systems work.

While I was in New York, I picked up skydiving as a hobby. Weekends at the drop zone. Nothing to do with work. Just a thing I did because it looked fun and because — well, you never know when you're going to need a skill.

That sentence is going to come up a lot in this book. It's maybe the most important thing I can teach you.

Years later, a company called iFLY was building indoor vertical wind tunnels and needed engineers who understood two things simultaneously: how mechanical and cooling systems work, and how human bodies behave in high-speed vertical airflow. The first part came from my engineering career. The second part came from throwing myself out of airplanes on weekends.

The combination got me the job. The combination put me in Qatar. The combination put me inside the world's largest wind tunnel on the day it first ran. A Saturday hobby and a pen clip. That's the honest version of how careers actually work.

• • •

This book is about how you build the kind of career where moments like that are possible.

It's not a book about AI, although AI is going to be in almost every chapter. It's a book about engineering judgment

— what it is, how you build it, and how AI changes the game for people who have it.

Because right now, we're in the middle of the biggest shift the engineering profession has seen since CAD replaced the drafting board. AI tools can generate calculations, write code, run simulations, produce reports, and draft designs at speeds that were unimaginable five years ago. This is genuinely transformative, and if you're starting your career right now, you have access to power that engineers my age didn't get until we were well into our careers — or ever.

But here's the thing that most people telling you about AI won't say: AI makes you faster, not smarter. It is the most powerful accelerant your career will ever see, but an accelerant on a bad process just gets you to a bad answers faster. The engineer who can produce confident-looking garbage at incredible velocity is the most dangerous professional in the industry. The engineer who has built real judgment — and uses AI to deploy that judgment at speed — is the most valuable.

The question isn't "how do I use AI?" The question is: how do I become the person whose judgment makes AI useful?

That's what this book answers.

I'm going to teach you how I think about problems, how I use tools, how I built the judgment that let me catch a four-degree error that would have cost a client hundreds of thousands of dollars, and how I went from a guy who thought AI was an annoying lie machine to someone who designs entire product lines in 48 hours with an AI team.

I'm going to tell you about the mentor who gave me the bedrock of my career in two and a half years, and about the project manager who nearly tanked a multimillion-dollar build because he wouldn't shut up and listen. I'm going to take you into the New York Public Library, where I spent a year and a half tracing 100-year-old piping through walls and ceilings, and into a Saturday afternoon where I accidentally became a CFD engineer because I said "fuck it" and installed free software I didn't know how to use.

And I'm going to tell you where this is all going. Because the future of engineering isn't 150-person firms doing 25 projects a year. It's teams of one or two people running 30 to 100 AI agents, doing hundreds of millions in business, with judgment as the only thing that separates the ones who thrive from the ones who get washed out.

You're going to start building that future now. Whether you're sixteen and reading this in your bedroom, or twenty-two and sitting at your first desk, or twenty-eight and wondering why your career feels stuck — the framework is the same. Build your judgment. Build your tools. Build your team.

Let's get started.

CHAPTER 1: RULE NUMBER ONE

Before I teach you a single tool, a single framework, a single trick for using AI to do engineering work — we need to establish something. This is the one thing in this entire book that is not flexible, not contextual, and not open to interpretation.

You cannot blame AI.

Not once. Not ever. Not when the deadline is tight and the output looked right. Not when the client is angry and you need somewhere to point. Not when the calculation was wrong and the system doesn't work and everyone in the room wants to know what happened.

You own everything that leaves your desk. Everything. If you used AI to generate it, you used a tool. The tool didn't sign the drawing. The tool didn't submit the report. The tool didn't stamp the calculation. You did. Your name. Your bond. Your word. The responsibility flows up to you, always, regardless of what produced the work.

On my team, the first person to blame AI gets made an example of. The next person is fired.

That sounds harsh until you think about what blaming AI actually means. It means you're telling me — and your client, and the contractor, and the code official — that you sent out work you didn't verify. That you put your name on something you didn't understand. That you were the quality control layer and you weren't paying attention. In any other context, that's grounds for termination. AI doesn't change that. If anything, it raises the stakes, because AI produces work at a volume that creates more opportunities for uncaught errors than you've ever had before.

THE JUNIOR ENGINEER TEST

Here's a way to think about it that makes it simple.

Imagine you have a first-year engineer on your team. Smart, fast, eager. They've read every textbook. They can produce calculations and reports at an impressive pace. But they've never set foot on a jobsite, they occasionally mix up their air-side and water-side numbers, and every so often they'll confidently give you a result that's just wrong — not because they're bad, but because they're inexperienced and working from pattern recognition rather than deep understanding.

That's AI. That's exactly what AI is.

Now: if that first-year engineer produced a calculation that went out under your seal and the system failed — would you blame the first-year? Or would you blame yourself for not reviewing it?

You'd blame yourself. Every licensed engineer knows this. The seal means you verified the work. It doesn't mean you personally did every calculation. It means you checked it, you understood it, and you stand behind it.

AI gets the same standard. Not a lower one. Not a higher one. The same one. You are the reviewer. You are quality control. You are the throat to choke when something goes wrong. That's the job. AI doesn't change the job — it changes the speed at which the job happens.

WHY THIS IS CHAPTER ONE

Everything in this book that follows — every tool I teach you, every workflow I show you, every framework for using AI to do engineering work faster and better — is built on top of this principle. It's not decoration. It's the foundation.

When I teach you the sniff test in Chapter 4, it's because you need it to live up to Rule Number One. When I teach you cross-checking in Chapter 5, it's because you need a way to verify AI output before your name goes on it. When I teach you the Einstein Method for problem definition in Chapter 3, it's because properly defining the problem is how you prevent the kind of errors that AI introduces when left unsupervised.

The entire framework exists because the responsibility never leaves you. The judgment-building isn't optional professional development. It's survival equipment. You are building the skills to own your AI-generated work the way you'd own hand calculations — with confidence that comes from real verification, not from hoping the tool got it right.

THE DOCTOR ANALOGY

People will tell you AI is going to replace engineers. You'll hear it from tech bros who don't understand what engineers do. You'll hear it from nervous classmates. You might even hear it from engineers who are scared of the change.

It's wrong.

Nobody is arguing that AI should replace doctors. Not seriously. AI is transforming medicine — better imaging analysis, faster differential diagnosis, broader literature review — but the doctor still makes the call. The doctor still sees the patient. The doctor still owns the outcome. AI is amplifying doctors' ability to deliver judgment. It is not replacing the judgment.

Engineering is the same. AI will amplify your ability to deliver engineering judgment at scale and speed that previous generations couldn't imagine. It will not replace the judgment itself. The buildings still need to stand. The systems still need to work. The physics still doesn't care about your prompt.

You are not being replaced. You are being amplified. But amplification requires someone worth amplifying. That's what the rest of this book is about.

FAILING PROPERLY

Here's a truth that might surprise you: you will fail. Not might. Will. AI will produce something wrong that you'll miss. You'll make a call with incomplete information and it won't work out. You'll fry a PLC board because you moved wires around on a live system without pausing to think about where that 24 volts was going to land.

I know, because I've done that last one.

The board wasn't expensive. Nobody got hurt. But a channel was dead and we needed a replacement to keep the project moving. Here's what I did: I called our supply chain guy, told him exactly what happened, and asked if he could get me a new board fast. Then I owned it in the engineering weekly call — told the whole team what I did, why it happened, and what the lesson was. Pause before you move live wires. Add that verification step. Build that "do nothing" moment into your process so the next person doesn't fry a channel.

That's the template, and it works for everything from fried PLC boards to missed calculations to AI-generated errors you didn't catch in time:

Own what happened. Explain the reasoning behind the call you made with the information you had. Don't blame anyone — not a person, not a tool, not circumstances. State the fix. State the process improvement. Execute.

Then move on. Living in the "whodunit" can feel productive, but it's not. Your team doesn't want to watch you grovel. They want to see you own it, fix it, and get back to work. That's what earns respect. That's what earns trust. And I mean this literally — you can actually earn credibility by failing properly. The engineer who breaks something, owns it immediately, shares the lesson, and adds a verification step to prevent it next time is an engineer people want on their team. The one who hides mistakes or deflects blame is a liability, regardless of how technically brilliant they are.

A project manager in France once told me something during my first wind tunnel commissioning that I've carried ever since: "If engineering school is supposed to teach you anything at all, it's that you know nothing. But it teaches you how to verify."

That's the deal. You know nothing — not with certainty, not without checking. But you know how to verify. That's your professional obligation. That's your value. And in the age of AI, where the volume of work you can produce has exploded, the verification is more important than ever.

Own it. Verify it. Put your name on it.

That's Rule Number One.

CHAPTER 2: THE PRETENDER MINDSET

There was a TV show in the late '90s called *The Pretender*. The premise was simple: a genius named Jarod who had studied so many disciplines and absorbed so much knowledge that he could walk into almost any situation — a hospital, a law firm, a racetrack, a crime scene — and convincingly operate as if he belonged there. He'd solve the problem, save the day, and disappear. The show was fiction, but the underlying idea is real, and it's the most important mental model I can give you for a career in engineering.

You are building The Pretender's skill set. Starting now.

That doesn't mean you're faking it. It means you're building such a dense, interconnected web of knowledge — physics, materials, codes, field experience, system behavior, failure modes, human factors, business economics, communication — that when a problem lands on your desk, no matter how unfamiliar it seems on the surface, you have something in your internal model to anchor to. You have context. You have adjacent experience. You have a starting point.

The engineers who will thrive in the AI age are not the ones who know the most about AI. They're the ones who've built the richest internal model of how the physical world works — and then use AI to operate on that model at speeds that were previously impossible.

YOU NEVER KNOW WHEN YOU'RE GOING TO NEED THIS

I want you to adopt a phrase as a personal mantra. Say it to yourself when you're sitting in a lecture that seems boring. Say it when you're on a jobsite survey and the task feels routine. Say it when someone is explaining something that doesn't seem relevant to your current project.

You never know when you're going to need this.

I picked up skydiving while I was designing buildings in New York City. It was a weekend hobby. Nothing to do with mechanical engineering. I learned to fly my body in freefall because it was fun and the people at the drop zone were interesting and it scared me in a way that felt worth leaning into.

A few years later, I was commissioning vertical wind tunnels — machines that make humans fly indoors. The job required someone who understood both mechanical and cooling systems AND how human bodies behave in high-speed vertical airflow. We called it the "human-ometer." Using your body to fly the air and feel for discontinuities that meant the system had leaks or rough spots that were causing unintended turbulence. QC with your body. My engineering career gave me the first part. My weekend hobby gave me the second part. The combination put me in a niche that almost nobody else could fill, and that niche became the defining chapter of my career so far.

Nobody plans these connections. You can't sit down at eighteen and draw a line from "skydiving hobby" to "vertical wind tunnel commissioning engineer." The line only makes sense looking backward. But the reason the line existed at all is because I was always in acquisition mode. Always absorbing. Always treating new knowledge and new experiences as future tools, even when I couldn't see the future they'd be useful in.

This is a mindset, not a study technique. It changes how you operate in every setting.

When you're in class, you're in class. Not on your phone. Not half-listening while you scroll. The lecture on thermodynamic cycles that seems boring today is the intuition that saves you at thirty when you're sizing a chiller plant and the numbers from AI "look right" but something in your gut says they're off. You won't be able to articulate why the numbers feel wrong unless you absorbed the fundamentals well enough for them to become instinct. That instinct is built in the boring lectures, not the exciting ones.

When you're in the field, you're absorbing. How does piping actually fit together? What does a properly installed duct look like versus a hack job? What does a pump sound like when it's cavitating versus when it's running clean? What does the mechanical room smell like when a bearing is going bad? These are sensory memories, and they become your calibration instruments for the rest of your career. You can't get them from a screen.

When you're in conversation, you're learning. The facilities manager explaining why he hates a certain brand of valve. The contractor telling you about a material substitution that worked on another project. The client describing a problem they've been living with for years. Every conversation with someone who has different experience than you is an opportunity to add to your model.

When you're off the clock, you're still building. Hobbies aren't distractions from your career — they're potential future differentiators that you can't predict yet. Skydiving made me a wind tunnel engineer. Woodworking teaches you about materials and tolerances and the gap between design intent and physical reality. Programming as a hobby teaches you how to think in systems. Cooking teaches you about heat transfer and process timing. Everything connects if you let it.

SKILL ACQUISITION IS SURVIVAL

Here's a framing that will put your brain in the right state: treat skill acquisition like a covert agent treats preparation.

You don't know what your next mission is going to be. You don't know if you'll need to understand controls wiring, or structural loading, or psychrometrics, or contract law, or supply chain logistics. You don't know which random Tuesday afternoon conversation is going to give you the one piece of context that makes a future problem solvable. So you treat everything as potentially mission-critical.

This isn't exhausting — it's energizing, once you adopt it. It turns boring situations into interesting ones. The meeting about code updates that everyone else zones out of? You're taking mental notes because you never know when a code interpretation is going to determine whether your design works or costs the client an extra hundred thousand dollars. The site visit where everyone else is checking their email while the contractor talks? You're asking questions because the contractor has thirty years of field knowledge and he's giving it away for free to anyone who bothers to listen.

This puts your brain in a state of focused learning and connection-making. It builds the internal universe that you'll use for judgment calls for the rest of your career. It's the compound interest of paying attention, and it starts returning dividends faster than you'd expect.

THE MODEL IN ACTION: SUBMARINES TO HVAC

My mentor Mike — the guy I'll tell you more about in Chapter 4 — designed submarines after college. That was his first job. Submarine thermal systems, environmental controls, the engineering of keeping humans alive in a sealed metal tube underwater.

Then he went to work at JB&B Engineering in Manhattan, designing building mechanical systems. Boilers. Chillers. Air handling units. Ductwork. The kind of stuff that keeps a building comfortable instead of keeping a submarine crew alive.

On the surface, those seem like very different jobs. Under the surface, the physics is identical. Heat transfer. Fluid dynamics. System control. Environmental management within a sealed or semi-sealed envelope. The principles Mike learned designing submarine life support systems translated directly into the judgment he brought to building design — and that cross-pollination made him better at building design than someone who had only ever done building design.

Mike's career is the Pretender Mindset across decades. He didn't plan the submarine-to-HVAC pipeline. He just built broad, deep knowledge in one domain, and when he moved to a different domain, the connections were already there, waiting to be useful.

You'll see this pattern in every successful engineer I've met. The structural engineer who also understands mechanical loads makes better structural decisions. The electrical engineer who understands the construction sequence designs more buildable systems. The HVAC designer who has spent time on jobsites designs mechanical rooms that contractors can actually install, instead of beautiful drawings that are physically impossible to build.

Breadth isn't the enemy of depth. It's the foundation that makes depth useful.

WHAT THIS MEANS IN THE AI AGE

Here's why this matters now more than ever.

AI can look up any fact. AI can recall any standard. AI can produce calculations in any discipline you point it at. If your only value is knowing things — storing information and retrieving it on demand — you're competing with a tool that does that infinitely faster than you.

But AI cannot build the interconnected model. It can't combine the sensory memory of a cavitating pump with the intuition from a thermodynamics lecture with the field observation that the piping layout doesn't match the drawings with the contractor's offhand comment that "we had to reroute because of the beam." AI can process all of those data points individually if you feed them in. It cannot connect them on its own in the way that produces engineering judgment.

Your internal model — the dense web of knowledge, experience, sensory memory, and pattern recognition that you build through the Pretender Mindset — is the one thing you bring to the table that AI doesn't have and can't replicate. It's what makes your judgment valuable. It's what lets you look at AI output and say "that doesn't smell right" before you can even articulate why.

AI is the engine. Your internal model is the driver. An engine without a driver goes nowhere useful. A driver without an engine walks. Put them together and you're moving at a speed and with a capability that neither could achieve alone.

But you have to build the model first. Or rather, you have to build it continuously, because it's never finished. Every day you're in class, in the field, in conversation, in new experiences — you're adding to the model. The richer it gets, the more valuable your judgment becomes, and the more powerful your AI-augmented output is.

So start now. Pay attention in the boring lecture. Ask questions on the jobsite. Pick up the hobby that has nothing to do with your major. Talk to the old guy in the mechanical room who's been keeping the building alive for forty years.

You never know when you're going to need it.

CHAPTER 3: THE EINSTEIN METHOD

Einstein reportedly said that if he had an hour to solve a problem, he'd spend fifty-five minutes defining the problem and five minutes solving it. Whether or not he actually said it, the ratio is right. And in the age of AI, the ratio has shifted even further — because AI has collapsed the "solving" phase to near-zero for many problems, which means the definition phase is where all the value lives.

Most engineering failures aren't failures of calculation. They're failures of problem definition. The wrong system gets sized perfectly. The right system gets designed for the wrong constraints. A beautiful analysis answers a question nobody needed to ask. The math is correct and the answer is still wrong, because the problem was never properly stated.

I've built my entire career on this idea, and I'm going to give you the method.

STEP 1: STATE THE PROBLEM IN PLAIN LANGUAGE

Not in engineering jargon. Not in the format your software wants. Not in the language of the spec or the RFP. Write it out like you're explaining it to a sharp twelve-year-old. If you can't do this, you don't understand the problem yet.

This sounds simple and it isn't. Engineers are trained to jump into technical framing immediately. Somebody says "the building is too hot" and your brain goes straight to cooling load calculations, chiller capacity, airflow rates. Stop. Back up. The problem is: people in this building are uncomfortable. Why? Maybe the cooling system is undersized. Maybe it's sized correctly but the controls are wrong. Maybe the building envelope is leaking air. Maybe the occupants changed the use of the space and now there are twice as many people generating heat as the original design accounted for. Maybe the thermostat is in the wrong location and the system thinks the space is satisfied when it isn't.

You haven't solved anything yet. You've just resisted the urge to jump to the first technical explanation, and in doing so you've opened up a problem space that's much more likely to contain the actual answer.

STEP 2: MAP THE CONSTRAINTS – THEN CHALLENGE THEM

List everything you think is fixed. Budget. Space. Codes. Timeline. Existing equipment. Client preferences. Material availability. The stuff that feels like the walls of the room you're working in.

Now go through each one and ask: *if I could eliminate this constraint, what options would that unlock?*

This is where breakthroughs live. A shocking number of "constraints" are actually assumptions that nobody has questioned. The budget was set by someone who didn't understand the scope. The space limitation is based on an architectural drawing that hasn't been updated. The timeline is driven by a deadline that turns out to be flexible if you ask the right person. The material specification is carried over from a previous project and doesn't apply here.

I'm not saying ignore real constraints. Code requirements are real. Physics is real. Structural capacity is real. But within any problem, there are hard constraints and soft constraints, and the soft ones are often where the biggest design improvements hide. Your job is to know the difference — and to have the courage to ask "is this actually fixed, or is this something we chose?"

STEP 3: IDENTIFY WHAT YOU DON'T KNOW

This is the hardest step and the most important one.

Young engineers are trained to project confidence. In school, you got graded on having answers, not on articulating questions. In the workplace, there's pressure to look like you know what you're doing, especially when you're new and surrounded by people with more experience.

Fight that pressure. The real skill is admitting — clearly, specifically, with precision — what you don't know.

Not "I'm not sure about this." That's useless. Instead: "I don't know the frost depth at this site. I don't know if the existing electrical panel has spare capacity. I don't know whether the client's stated budget includes demolition. I don't know the actual occupancy schedule for this space."

Each unknown is one of two things: something you need to go find out, or a risk you're choosing to carry forward. Both are valid. But you have to identify them explicitly. The unknowns you don't name are the ones that kill projects.

A project manager in Lyon, France told me something during my first wind tunnel commissioning that I've carried for my entire career: "If engineering school is supposed to teach you anything at all, it's that you know nothing. But it teaches you how to verify."

Your job is verification. Not confidence. Not certainty. Verification. And you can only verify things you've first admitted you don't know.

STEP 4: IF THE PROBLEM IS PROPERLY DEFINED, THE ANSWER IS OBVIOUS

This sounds like a bold claim. It is. And experienced engineers know it's true.

When you truly understand the constraints — the real ones, not the assumed ones — and you've identified what you don't know, and you've stated the problem in plain language that strips away the technical clutter... the solution space narrows dramatically. You're not choosing from infinite options. You're choosing from two or three, and the tradeoffs between them are clear.

I designed cooling systems for iFLY for almost ten years. These are indoor vertical wind tunnels — machines that generate enormous, rapidly fluctuating heat loads inside a basically closed-loop system. Every time I made one of those systems work, the people around me acted like it was voodoo. Like I had some mystical ability to tame a uniquely complex problem.

I didn't. I had a process.

Start with the physics: energy in equals energy out. Model the load. Understand the rate of change — the tunnel's heat load swings fast, so the cooling system needs to react fast. That means faster chiller response and fast-acting valve modulation to handle quick transients.

Once I had the load data, I didn't need anyone else. I had a repeatable method that I'd walk through the numbers: load profile to chiller selection to controls theory to efficiency analysis. I could build systems that were cheap upfront and costlier to own, or costlier to buy and cheaper to own, depending on what the client needed. Same method, different constraint emphasis, different output.

People would watch me move a mountain in a few days and think I was walking on water. I just had process. I had just defined the problem properly.

And every time someone new tried to reinvent the approach — adding unnecessary complexity, rethinking the controls strategy, trying to account for conditions that didn't actually affect a closed loop — they'd blow the budget and the timeline. I'd be sitting in the meeting saying "I've run those traps. They bite you here and here. We do this because it balances cost and performance and we've iterated on it." After about ten iterations over a decade, if you haven't gotten close to optimal, you're either a bad engineer or a bad learner. We had the system dialed.

That's the Einstein Method in practice. Define the problem thoroughly — the physics, the constraints, the unknowns — and the solution becomes obvious. Not easy. Obvious. There's a difference. But obvious is ninety percent of the battle.

THE RUN READY RITUAL

There's a companion to the Einstein Method that lives at the other end of the timeline — not when you're defining the problem, but when you're about to act on the solution.

The easiest thing you can ever do is do nothing.

Don't push the button. Don't send the command. Don't change the setpoint. Don't start the system. Just... pause.

The moments right before action are your last chance to check yourself. I've made this into a ritual. Before every test run on a wind tunnel, I would ask every single person on the team: "Are we run ready?"

Not a rhetorical question. Not a formality. Every person was asked, individually, and every person was expected to give a real answer. Because you don't know who knows the gotcha.

The answers I've gotten from that simple question have prevented hundreds of mistakes:

"There's debris in the flow path."

"Chiller system isn't ready — you have limited heat rejection capacity."

"Tony needs five more minutes. He's mid-update on the parameters."

Every one of those is a potential failure, safety incident, or wasted test run that was caught by a thirty-second ritual. No single person held all of that information. The only way to surface it was to ask everyone.

This applies to every engineering action, not just wind tunnel testing. Before you submit the report, before you send the calculation, before you finalize the design — pause. Do a quick sanity check. Sniff test the numbers. Ask someone to glance at it. Give yourself one more beat to catch the thing your subconscious has been nagging you about. You're at the end. You don't have any deadlines pressing you anymore. Take this calm moment where you feel good to check your work for sanity. Small edits can make big differences and you're in a great head space to do it because you're at the end, rather than still in the pressure cooker.

The number of times I've paused before a startup and thought "actually, I should probably close that valve or restrict that parameter first" is in the hundreds. Each one of those pauses cost me ten seconds and potentially saved hours of rework or worse.

120 DEGREES IN THE DESERT – A STORY ABOUT CHECKING YOUR INPUTS

I want to tell you a story about what happens when the problem definition is wrong and what it looks like when trained judgment catches it.

I was consulting on a cooling system for a vertical wind tunnel in one of the hottest areas in the United States. The system was underperforming, and another engineering firm had produced a report diagnosing the issue. Their diagnosis was correct: the chillers had been improperly selected for the actual operating conditions.

The report was good work. Thorough analysis, clear writing, correct conclusions about the selection criteria. It also stated that the local NOAA climate data showed temperatures at or above 120°F for 0.4% of days in recent history.

At 120°F ambient, refrigeration technology hits a wall. The temperature difference between your coil surfaces and the outdoor air becomes so small that you need enormous surface area to reject heat. A chiller that's nominally rated at 550 tons will struggle to make 300 tons at those conditions. The capacity falls off a cliff.

We pulled selections confirming this. The numbers said we'd need two maximum-frame chillers to meet the load. Big equipment. Big money.

I was on a call with the customer, walking them through the selections. "The report referenced local NOAA data over 120, so that's where we're specifying the equipment for..."

And then my brain stopped me. Mid-sentence. Right there.

The thought was: *These are the same guys who called me for help understanding how this cooling system works in the first place. Why would I trust their climate data without checking it myself?*

I stopped talking and started pulling up the ASHRAE design conditions for the area. The design temperature wasn't 120°F. It was 116°F.

Four degrees. In normal circumstances, four degrees is a rounding error. But when you're operating at the absolute edge of a chiller's performance envelope — where coil surface temperatures are 124 to 126°F and you're trying to squeeze every last degree of temperature difference out of the system — four degrees is enormous.

When we reran the selections at 116°F, each chiller dropped about 20% in frame size. The entire project economics changed. Not because the engineering methodology was different, but because the single input that everything else was built on — the design temperature — was wrong.

The report was otherwise correct. That's what made it dangerous. When a document is ninety-five percent right, your guard drops on the other five percent. You read through it, you nod along, you accept the inputs because everything else checks out. The one wrong number hides in the shadow of all the right ones.

Correct reports are the most dangerous place for wrong numbers to hide.

WHERE AI FITS IN PROBLEM DEFINITION

Here's where the Einstein Method and AI become a genuinely powerful combination.

AI is a spectacular tool for problem definition — arguably better here than anywhere else in the engineering workflow. You can use it to rapidly research codes, standards, and precedent projects. You can explore "what if" scenarios when challenging constraints: "What happens to system cost if I switch from glycol to water? What if we run the CFD from the other angle? What if the occupancy schedule shifts by two hours?" You can use it to visualize data in ways that expose patterns your spreadsheet is hiding or it would take too much time to parse out. You can build quick webapps and dashboards that let you see the problem space from different angles. You can run fast parametric analyses to understand which variables the solution is most sensitive to.

AI makes the definition phase faster and more thorough, which is exactly where you want to invest that speed. Fast definition on a complex problem is engineering at its best. You're exploring the solution space, testing assumptions, identifying sensitivities — all before you commit to a direction.

But the judgment — which constraints to challenge, which unknowns actually matter, whether the problem statement is complete, whether the inputs are correct — that stays with you. AI didn't stop me mid-sentence on that call. My trained instinct did. AI didn't tell me to question the 120°F figure. Twenty years of working at the edge of chiller performance envelopes did.

Define the problem thoroughly. Challenge what you think is fixed. Name what you don't know. Verify your inputs from primary sources. And then — when the answer becomes obvious — pause one more time before you act on it.

Are we run ready?

CHAPTER 4: CALIBRATING YOUR INSTRUMENTS

Here's the uncomfortable truth for young engineers entering the profession right now: the AI age demands more judgment than ever, and you have to build it faster than any previous generation.

Your predecessors had ten years of apprenticeship. They sat next to senior engineers, watched how decisions were made, absorbed rules of thumb through osmosis, and gradually built the internal calibration that let them look at a number and say "that's wrong" before they could explain why. The timeline was slow, but the method worked. Judgment accumulated through thousands of hours of supervised repetition.

You don't have that luxury. Within two or three years of starting your career, you'll be expected to run AI-assisted workflows that can produce building designs, equipment selections, and engineering reports at industrial scale. The output will look polished. The formatting will be professional. The calculations will be neatly organized. And some of it will be wrong — not obviously wrong, but subtly wrong in ways that only calibrated judgment can catch.

So how do you calibrate your instruments in compressed time?

ALWAYS ASK WHY

This is the single most powerful calibration tool you have, and it costs nothing.

Why does that duct fitting use a long-radius elbow instead of a short-radius? Why do we accept a higher pressure drop in this part of the system but not over there? Why is the economizer lockout set at 55°F and not 60°F? Why is the pipe oversized here when the flow rate doesn't seem to justify it? Why did the designer put a heat exchanger between these two loops instead of connecting them directly?

Every "why" you chase down deposits another data point in your internal model. The engineer who asks why a hundred times has a richer judgment base than the engineer who memorized a hundred answers. Answers are static. The reasoning behind answers is dynamic — it connects to other reasoning, it builds frameworks, it creates the pattern recognition that eventually becomes instinct.

When you look at a drawing, a system, a piece of equipment — interrogate it. Don't just note what's there. Ask why it's there, why it's sized that way, why it's in that location, why it's that material. The answers build your calibration faster than any textbook, because they come with context that textbooks can't provide.

The people around you are going to get tired of you asking why. That's fine. The good ones will recognize what you're doing and make time for it. The bad ones will tell you to just use the number and stop asking questions. That difference — how your boss responds to "why" — is one of the best indicators of whether you're in the right environment. More on that later in this chapter.

THE SNIFF TEST

The sniff test is shorthand for: does this result smell right?

Same as testing the leftovers in your refrigerator. You open the container, take a sniff, and you know instantly — either it's fine or it's going straight in the trash. You don't need to run a lab test. You don't need to check the expiration date. Your nose knows.

In engineering, it's the moment you look at an AI-generated equipment schedule and something feels off about a chiller being 200 tons for a 15,000-square-foot building. It's the moment a CFD visualization shows airflow patterns that don't match what you've seen in real spaces. It's the moment a calculation comes back with a pipe size that your gut says is too small, even though the math appears to check out.

This isn't mysticism. It's pattern recognition built through exposure. And there are specific practices that build it faster.

Rules of thumb as guardrails. Memorize the fundamental rules of thumb for your discipline. I mean actually memorize them — commit them to the same place in your brain where you keep your phone number and your home address. Watts per square foot for different building types. Velocity limits in ductwork. Pressure drops per hundred feet of pipe. CFM per ton for comfort cooling. These aren't precise numbers. They're smoke detectors. When AI output violates a rule of thumb, it doesn't necessarily mean the AI is wrong. But it means you need to understand why the number is different from what you'd expect. The investigation that follows is where the real learning happens.

Cross-checking obsession. Make it a personal standard: nothing leaves your desk without being verified by a different method than the one that produced it. AI-generated calculation? Check the critical numbers with a hand calc or a different tool. Hand calculation? Spot-check it with AI. Two independent methods agreeing builds confidence. Two independent methods disagreeing demands investigation. This is peer review applied to your own work, and it should become automatic — like checking your mirrors before changing lanes.

Deliberate failure practice. This one sounds counterintuitive, but it's one of the fastest ways to build calibration. Use simulation environments to intentionally break things. Oversize a pump by ten times and see what the model tells you. Remove the insulation from a pipe in your energy model and watch what happens to the system performance. Double the occupancy in a space and see how the cooling load responds. These exercises build pattern recognition for what "wrong" looks like — and more importantly, they force you to ask why the failure presents the way it does. When you've seen what a grossly oversized pump does to a system in simulation, you'll recognize the early signs of it in real equipment selections.

THE EYE-CROMETER

The eye-crometer is your visual calibration — the ability to look at a drawing, a model, a rendering, a data plot, equipment alignment, and immediately sense whether it's in the right ballpark.

This develops through exposure, and there's no shortcut. Look at hundreds of mechanical plans. Study thousands of data plots. Review scores of equipment submittals. Over time, your eye becomes a remarkably accurate instrument for detecting anomalies. A duct that looks too big for the space. A pump curve that doesn't match the system curve in a way that suggests the operating point is wrong. A piping layout that looks constructable on paper but physically can't fit when you account for insulation thickness and maintenance clearances.

AI can actually accelerate your eye-crometer development by generating more scenarios for you to review. Use it to produce variations on a design — what does this system look like at half capacity? At double? In a different climate zone? With a different equipment selection? Each variation is a calibration point. The more you see, the faster your visual instinct develops.

But the foundation of the eye-crometer is physical. You have to see real systems, real installations, real equipment in real spaces. The engineer who has walked through a hundred mechanical rooms will look at a drawing and immediately notice that the six-inch valve shown next to the pump can't physically fit in the clearance available. The engineer who has only seen drawings won't catch it until the contractor calls from the field with a problem.

LEARNING TO READ AT ENGINEERING SPEED

The eye-crometer is your visual instrument. You also need a textual one — the ability to rip through technical documents at high speed with real comprehension. Not skimming. Not getting the gist. Actually understanding what a hundred-page manual is telling you well enough to act on it.

This is a learnable skill, and it's one of the biggest velocity multipliers you can develop. When you can absorb a mechanical specification, an equipment manual, an ASHRAE chapter, or a code section in minutes instead of hours, you operate at a fundamentally different pace than the engineer who has to grind through every page at the same speed.

Here's what's actually happening when an experienced engineer reads fast: they're not reading every word. They're not using some trick to bypass comprehension. They're leveraging domain knowledge to process text in chunks rather than individual words. When you know what "variable frequency drive" means at a bone-deep level, your brain doesn't read three words — it reads one concept. When you've seen fifty equipment submittals, the sixty-first is mostly familiar territory with a few novel details. Your eyes learn to find the novel details and skip the boilerplate. The speed comes from expertise, not from bypassing the reading process.

But here's the part most people miss: you can deliberately accelerate how fast this develops. There's a method to it, and the method is practicable from day one — even before you have deep domain knowledge.

Step one: stop saying the words in your head. This is the entry-level skill and it's the single biggest unlock. Most people read by looking at a word, saying it silently in their internal monologue, and then understanding it. Look → pronounce → comprehend. Three steps. That middle step — the silent pronunciation — is an anchor dragging behind you. It limits your reading speed to your speaking speed, which tops out around two or three hundred words per minute. That's a hard ceiling, and no amount of strategy will break through it while that habit is in place.

Here's the shift: words aren't sounds. They're glyphs. They have meaning that your brain can access directly, without the detour through pronunciation. You need to go from look → comprehend. Skip the phonetics entirely. When you see "heat exchanger," you don't need to hear it in your head to know what it means. Your brain already has the concept loaded. Let your eyes deliver meaning directly.

This will feel strange at first because school teaches us to read out loud, then to read silently — but "reading silently" for most people is still reading out loud, just with the volume turned off. The internal monologue is still running. Turning it off takes deliberate practice. Start by pushing your eyes faster than your internal voice can keep up. Force the pace. Your brain will resist for a while, then it'll adapt — it'll start grabbing meaning from the shapes on the page without waiting for the voice to catch up. Once that clicks, you've removed the speed limit. Everything else builds on top of this.

Step two: know why you're reading before you open the document. This sounds obvious, but most people skip it. Are you reading this manual to understand the full system? To find one specific operating parameter? To check whether the equipment meets your spec? Different purposes require different reading depths, and if you don't set the purpose first, you'll read everything at the same plodding speed because your brain doesn't know what to prioritize.

Step three: preview the structure before you read the content. Spend two or three minutes scanning the table of contents, the section headings, the diagrams, the summary tables. Build a mental map of how the document is organized. This does two things — it tells your brain where the information you need lives, and it gives you a framework to hang the details on when you encounter them. Reading without structure is like navigating without a map. You get there eventually, but you take every wrong turn along the way.

Step four: read at variable speed. This is the core strategic habit. Don't read every section at the same pace. Blow through the sections you already understand — the standard boilerplate, the general descriptions, the material you've seen in twenty previous manuals. Slow down hard on the sections that are new, unusual, or directly relevant to your purpose. Your reading speed should fluctuate constantly — fast on familiar ground, slow on unfamiliar ground, stopped completely when something doesn't match your expectations.

That last part is the sniff test applied to reading. When you're moving through a document at speed and something snags your attention — a number that doesn't look right, a configuration you haven't seen before, a claim that contradicts your experience — that's your brain telling you to slow down. Listen to it. The ability to detect those snags while reading fast is the whole skill.

Step five: make multiple passes when the document matters. First pass for structure and overview. Second pass for the sections that are critical to your purpose. Third pass — if needed — for the details you'll be putting your name on. Each pass is faster than the last because you already know the landscape. This feels slower than reading once at medium speed, but it produces better comprehension in less total time because each pass is targeted rather than exhaustive.

Step six: practice on volume. Read more. Read everything. Equipment catalogs, specification sheets, code sections, trade publications, competitor literature, installation manuals for equipment you'll never specify. Every document you read in your domain deposits another layer of familiarity that makes the next document faster. This is the same compounding that builds the sniff test and the eyecrometer — volume creates pattern recognition, and pattern recognition creates speed.

The honest truth is that there's no shortcut past the reading hours. The speed comes from knowledge, and knowledge comes from volume. But the method — kill the internal monologue, set your purpose, preview the structure, vary your speed, make multiple passes — turns those hours into a much steeper learning curve than grinding through documents cover to cover at the same pace. Start practicing these habits now, with whatever you're reading for school or work. Within a few months, you'll notice you're moving through material significantly faster. Within a year or two, you'll be pulling the critical information out of thick documents at a pace that seems almost unreasonable to the people around you.

It's not magic. It's calibration. Same as everything else in this chapter.

GO LOOK AT THINGS

No amount of simulation replaces standing in front of the real thing.

This deserves emphasis because the AI age creates a seductive trap. Everything can be modeled, visualized, and analyzed from your desk. You can run CFD simulations, generate 3D renderings, pull manufacturer data, and produce complete engineering packages without ever leaving your chair. The temptation to stay at the screen is enormous. The tools are so good now that it feels like you're seeing the real thing.

You're not. The screen shows you what the model contains. The field shows you what the model missed.

Go into the field. Go into the mechanical room. Go to the jobsite during construction. Go to the factory where your equipment is manufactured. Every time you have the opportunity to see the physical reality behind your digital models, take it.

But here's the part most young engineers miss: being there isn't enough. You have to be an aggressive participant in your own education.

The field is full of people who know things you don't, and many of them didn't learn it from a textbook. The pipefitter who's been bending copper for thirty years knows things about how systems actually go together that no simulation will teach you. The controls technician who's commissioned two hundred air handlers has intuition about startup sequences that your professor never covered. The facilities manager who's been keeping a building alive for two decades knows where every design decision went right and where it went wrong.

Your job is to recognize that experience in others and pull it out of them. Ask judgment questions, not just factual ones. Not "what size is that pipe?" but "why did you run it that way instead of the shorter route?" Not "what's that valve for?" but "what happens to the system when you close it?" Not "what's the setpoint?" but "how did you arrive at that number, and what happens when the load swings?"

People who've spent decades doing physical work are often deeply generous with their knowledge when they encounter a young engineer who's genuinely curious rather than just checking a box on a site visit. They can tell the difference instantly. The engineer who walks through a mechanical room with their hands in their pockets and their eyes on their phone is invisible to them. The one who points at things, asks questions, and actually listens to the answers gets the thirty-year download.

The engineers who will lead in the AI age are the ones who pair digital speed with physical grounding. You can run a thousand simulations in a day, but the engineer who has also stood in the space, heard the noise, felt the vibration, talked to the person who maintains the system — that engineer will make better decisions about what those simulations mean.

EVERYTHING IS THERE FOR A REASON

Here's a principle that will transform how you see built systems: nothing you're looking at is accidental.

Every valve, every heat exchanger, every bypass line, every piece of equipment in a mechanical room represents a decision someone made — and more importantly, money someone spent. Owners don't buy things they don't need. That valve didn't appear for decoration. It isn't there to get in your way while you're climbing around the equipment. It's there to isolate a system, enable maintenance, protect something, or solve a problem that was painful enough to justify the cost.

This means that before you change a system — before you remove a component, reroute piping, or simplify a design — your first obligation is to understand why everything that exists was put there in the first place.

That heat exchanger separating two loops? It introduces energy losses. It costs money to buy, install, and maintain. So why did someone accept all of that? Because there was something going on — different pressures, different fluids, different ownership boundaries, contamination risk — that made the separation worth paying for. If you remove it without understanding the original reasoning, you may re-introduce the exact problem it was solving.

This is one of the richest learning environments you'll ever have, and it's available to you every time you do a field survey — or frankly, every time you walk into a public building or walk through your own house. Engineered systems are everywhere. You're surrounded by design decisions. When you're alone in a mechanical room, not being pressed by your boss to get back to your desk, that's your classroom. Point at things. Trace piping. Follow the flow. Ask yourself: why is this here? What problem does it solve? What would happen if it wasn't here?

Use your phone to photograph things you can't explain and research them later. This is a place where AI becomes a great study partner — describe the configuration you saw and ask what engineering problem that arrangement typically solves. You'll get answers that connect your physical observation to engineering principles, and that connection is how calibration happens.

Over time, this practice builds something invaluable: a library of design intent. Not just what systems look like, but why they were built that way. That library becomes the foundation of your judgment. When you're designing a new system and AI suggests an elegantly simplified configuration, your library of intent is what tells you "that looks clean, but the last time I saw a system without that isolation valve, it was because someone didn't understand why it was there in the first place — and they regretted it."

I worked at iFLY for almost ten years, building and refining cooling systems for vertical wind tunnels. By the time I left, those systems had been through about ten change iterations. Each iteration was filtered through two questions: can we eliminate this component and still get the same performance, and how do we make it more efficient for both capital and operating cost? The decisions we chose not to take are the biggest decisions we make. Every component that survived ten rounds of that questioning had earned its place.

But every time a new engineering team came on to one of our projects — especially the wholly-owned tunnels where an outside engineer was the Engineer of Record — they'd show up with "ideas" about how to improve the system. The conversation always went the same way. We've tried that. It doesn't work because it causes these issues. Here's the data from when we tried it.

Two types of engineers emerged from that conversation. The ones who started listening, recognized the depth of iteration behind the design, and began learning from it. And the ones who assumed they already had the competence, despite having zero experience in the application. They'd lost the plot of being an engineer. They went to verify — with me, my data, my experience — and then ignored what they found. You could be generous and say their sniff test was going off, that something about the application genuinely struck them as wrong. But a real sniff test violation demands investigation. You follow up. You check the reasoning independently. You verify before you override. They didn't do any of that. They just assumed they knew better.

The iFLY London project was the most expensive example. The outside engineer deviated from the proven design against our recommendations. Over fifty thousand pounds in rework when the system didn't perform. His first thought should have been simple: why are they doing all these things so oddly? Certainly these people aren't just slamming their heads against the wall over and over again. His next thought should have been to understand each component, verify my explanations by thinking them through independently, and be sure he understood the reasoning before making changes.

He didn't do any of that. And it cost everyone time, money, and frustration.

FIND YOUR MIKE

My second job out of college, I got hired at JB&B Engineering in New York City. One of the oldest and most respected engineering firms probably in the world. They had a particular way of operating. A steady supply of young, hungry engineers from colleges across New York, New Jersey, and Pennsylvania who wanted to break into engineering in New York. Above them were the mid-level associates, mostly experienced, average age skewing into the fifties with a few up-and-comers sprinkled in. Above them, the Partners. The best projects in the city. A lot of overtime available if you wanted it.

I was very lucky to get paired with a guy named Mike. Mike was a Jersey native who had joined JB&B as his second job, after designing submarines out of college. He'd been at the firm for at least twenty years. His accounts included the New York Stock Exchange, the New York Public Library, the International Coffee Exchange, Brooklyn Botanical Gardens, and a handful of early data center projects.

Mike was the single best teacher I've had in twenty years of engineering. In two and a half years, he gave me the bedrock of the rest of my career.

The reason was simple: he always made time for the "why." Why do we size it this way? Why is that the value you use and not this one? He'd always point me in the right direction and unlock my obstacles. It was never "because I said so" or "just use this number." It was always the reasoning, the origin, the engineering judgment behind the decision.

It was a two-way street. He mentored, and I got our projects locked down. I pushed, and he pulled. The exchange was: he gave me his accumulated judgment, and I gave him excellent work that freed him up to run his accounts. Both sides benefited. Both sides grew.

That's how real mentorship works. It's not charity. It's a transaction — and that's what makes it sustainable. You earn mentorship by being worth mentoring. Show up. Do the work. Make your mentor's life easier. Ask sharp questions that show you've been thinking, not lazy questions that show you haven't tried. Deliver quality that makes them trust you with more responsibility. The knowledge flows when the exchange is working.

After two and a half years with Mike, I made the mistake of leaving. Another company offered more money, and I figured it was time to test my new skills in the market. The new boss ripped me apart and tore me down in about a year. The next job after that? I walked on water again.

The variable wasn't my skill level. It was the environment. Same engineer, radically different outcomes based on who was above me and whether they were invested in my development or just interested in extracting labor.

Here's what I want you to take from that: if you're in your early twenties and your boss is a fuckin asshole, move. Don't stick around for the resume line. Don't stick around for the paycheck. Don't tell yourself you're "paying dues" or "building character." Your early career years are the highest-leverage learning period you'll ever have. Every month spent under a mentor who doesn't teach is a month of compounding judgment development that you'll never get back.

The billionaires of the world preach "hire fast, fire fast." The same goes for your boss. Find a good one, soak up the knowledge, and give them awesome work in return. When the environment stops being productive, move to one that is. Your career is an iterative design process — same as any engineering problem. Run the experiment, evaluate the results, adjust the parameters.

THE NEW YORK PUBLIC LIBRARY

One of Mike's accounts was the New York Public Library. Over a hundred years old. A protected landmark. And underneath its beautiful architecture was layer upon layer of mechanical, electrical, and plumbing systems that had been installed, modified, expanded, and partially abandoned across a century of mechanical innovation.

I spent about a year and a half mapping the piping in that building.

There were no comprehensive drawings. Most had been lost to time. A few drawings existed for later additions, but most of those were architectural only — they showed walls and spaces but not the systems running through them. We were literally redefining the problem from scratch, which meant someone had to trace every pipe, every duct, every system, through every wall and ceiling and floor in that building.

That someone was me.

So many dead ends. So many twists and turns through walls and over ceilings. I'd follow a pipe until it disappeared into a wall, then go upstairs and try to pick it up on the other side. Down to the basement. Back up to the fourth floor. Through rooms that hadn't been opened in years. I traced the condenser water pipes all the way from the mechanical room — where the original boiler used to live — through the old boiler flue, up to the roof where the cooling towers sat. The flue had been repurposed to carry condenser water after the city started providing steam through the street mains. The Library had switched their heating infrastructure over to city steam, which freed up the boiler room to become a mechanical room for the cooling equipment that came later when air conditioning became standard in public buildings.

I was reading the history of New York City's infrastructure in the walls of that building. Older steam systems with chilled water piped underneath. The order in which the city's utilities grew, visible in the layers of installation. When air conditioning was added. When systems were repurposed. When things were abandoned in place because it was cheaper to leave the dead pipe than to remove it.

I mapped every system and every dead pipe on hand sketches, brought them back to the office, and put them into CAD.

I also spent a lot of time in the office of the lead building engineer — a typical old-school greaser who knew every nut and bolt in the place and had at least forty years of history in that building. Whenever I lost a pipe in the walls, he'd help me figure out where it popped back out. He knew things about that building that no drawing could tell you, and he gave that knowledge freely to the young engineer who kept showing up and asking questions.

The payoff was enormous. Knowing where everything was made it straightforward to design the new systems. When I'd go back to survey with Mike, I could take him anywhere in the building and show him exactly where things went. He recognized pretty quickly that I'd cracked the code of the Library's systems, and he trusted me almost implicitly when it came to that project. I got sent solo to meetings. When I said that system was in room 223 in the ceiling, he trusted it and moved on.

I made myself an expert in that building. Not because someone assigned me to be the expert. Because I was in that low-pressure zone — no boss hovering, all the time in the world to look around — and I chose to chase every pipe to its end.

Why are there three different chiller types in this room? Why is it so hot in this room with the heat exchangers? Why does this pipe go up four floors and come right back down?

Each question was a thread, and each thread led to an answer that built my understanding of how real systems evolve over real time in real buildings. That understanding became judgment that I've carried into every project since.

The NYPL was my graduate school. The building was the textbook. The lead engineer was the professor. And the tuition was free — it just cost me the willingness to be curious and the persistence to follow every pipe until I understood where it went and why.

PUTTING IT ALL TOGETHER

Everything in this chapter connects. The "always ask why" habit feeds the sniff test. The sniff test feeds the eye-crometer. Field experience feeds all of them. Mentorship accelerates the whole process. And the principle that everything is there for a reason gives you the lens to learn from every system you encounter.

These aren't separate skills. They're one skill — engineering judgment — developed through multiple channels simultaneously. The engineer who asks why, checks their gut against rules of thumb, cross-checks with different methods, spends time in the field, learns from mentors, and studies existing systems with curiosity... that engineer is building judgment at maximum speed.

And in the AI age, that judgment is the whole game. AI produces output. You produce judgment about that output. The faster you calibrate your instruments, the sooner you become the person whose name on a drawing actually means something.

Start calibrating now. Ask why at every opportunity. Memorize your rules of thumb. Go look at things. Find your

Mike. Trace the piping. Follow every pipe until you know where it goes and why.

The instruments you're building today are the ones you'll rely on for the rest of your career.

CHAPTER 5: AI AS SPEED, NOT INTELLIGENCE

Here's the mental model I want you to carry for the rest of your career: AI is your deployment engine, not your brain.

It builds the spreadsheet. It sets up the CFD simulation. It generates the report. It writes the Python script to parse that equipment schedule. It drafts the submittal review. It writes code in programming languages you've never studied. It produces work product at a speed that would have seemed impossible five years ago. And all of that is genuinely, transformatively useful.

But it doesn't understand what it's doing.

That's not a criticism. It's a description of the mechanism, and understanding the mechanism is critical to using the tool well. So let me explain what's actually happening inside these systems, because if you're going to build your career on them, you need to know what you're standing on.

THE 1+1 PROBLEM

When AI tells you that 1+1 equals 2, it's not adding. It's predicting that "2" is the most probable next token in the sequence, based on patterns in its training data. It has seen "1+1=" followed by "2" so many times that "2" has an overwhelming probability of being the correct completion.

When a calculator tells you that 1+1 equals 2, it's actually performing the addition. It's running a deterministic operation that will produce the same answer every single time with the same inputs.

Both give you "2." But one is derived and the other is guessed. For 1+1, the guess is perfectly reliable. For the edge case of an unusual heat transfer coefficient, an atypical pressure drop in a system configuration you've never seen before, or a complex interaction between multiple variables at extreme operating conditions — the guess reflects the statistical center of the AI's training data, which may or may not reflect the physics of your specific situation.

This is the foundational distinction you need to internalize: there's a difference between a system that produces the right-shaped answer and a system that produces a right answer. For routine calculations where the AI has seen thousands of examples, the shape and the reality converge. For unusual conditions, novel configurations, or edge cases, they can diverge — and the divergence is invisible unless you have the judgment to catch it.

DETERMINISTIC VS. PROBABILISTIC – THE HARD LINE

Engineering requires determinism in most cases. When you put the same inputs into a pipe sizing calculation, a structural load analysis, or a heat transfer equation, you need the same answer every time. Buildings stand or fall on this. Systems perform or fail on this. Codes are written around this. Your professional seal means you verified the work through a reliable, reproducible method.

AI is probabilistic. Ask it the same engineering question twice and you may get two slightly different answers, both plausible, potentially only one (or zero) correct. The variation is usually small, but "usually small" is not an engineering standard of care.

The rule is simple and non-negotiable: **deterministic calculations live in deterministic tools.** Engineering software with locked formulas. Spreadsheets with verified equations. Hand calculations with known methods. Manufacturer selection software with published algorithms. These tools give you the same answer every time with the same inputs, and that reproducibility is what you put your name on.

AI is the deployment layer around those tools. It writes the spreadsheet. It configures the simulation. It sets up the selection software. It generates the report that wraps the results. It produces the deliverable package. But the numbers that determine whether a system is safe, whether it meets code, whether it will actually perform — those numbers come from deterministic processes. This will never, and should never, change

There are legitimate engineering uses for probabilistic methods. Monte Carlo simulations. Reliability analysis. Risk assessment. Weather-dependent energy modeling. But using probability in engineering requires more judgment, not less, and more safety guardrails — guardrails that are themselves based on good deterministic engineering judgment. You use probabilistic methods when you understand the distribution, the confidence interval, and the consequences of being in the tail. You don't use them because a chat window gave you a number and it looked right.

TRAP DOOR HALLUCINATIONS

Everyone who uses AI knows about hallucination — the tendency to make things up. You ask for a reference and it invents a paper that doesn't exist. You ask for a specification and it generates a plausible-sounding but fictional product number. These are annoying, but they're usually easy to catch because the fabrication is obvious when you check.

There's a more dangerous failure mode that I call the trap door hallucination.

This is when AI produces something that looks exactly like a real engineering rule of thumb. "Values for X are typically between 1 and 4.3." It "thinks" it with confidence. It uses it like authoritative knowledge. It even contextualizes it in a way that sounds like something a senior engineer would tell you.

But it's not a real rule of thumb. It's a pattern that emerged from training data — a statistical artifact of how often certain numbers appeared near certain terms. It has no engineering heritage. It wasn't derived from first principles. It wasn't validated against physical systems. It wasn't developed by an experienced engineer who understood the edge cases where it breaks down.

Real rules of thumb have heritage and boundaries. When a senior engineer tells you "figure about 400 CFM per ton for comfort cooling," that rule comes from somewhere — ASHRAE guidelines, decades of installed systems, thousands of hours of field commissioning. It also comes with invisible fine print that the senior engineer carries in their head: this works for standard commercial occupancies, it falls apart in high-density spaces, it doesn't apply to process cooling, and if you're above 7,000 feet of elevation the air density changes throw it off.

AI's invented rules carry none of that. There's no understanding underneath the number — just probability. No heritage, no caveats, no sense of where the rule breaks down. And because AI presents it exactly like a real rule of thumb, your sniff test has to be sharp enough to catch it. The results land close to correct — close enough to pass a casual glance. But "close enough" has a way of blowing up when your specific situation falls outside the range that the fabricated rule accidentally covered.

There's a reason I call this a trap door. You think you're walking on solid ground. The numbers look healthy. Not too far from what you expected. Everything seems fine — and then the floor falls out from under you. One of your conditions fell outside the range that the made-up rule of thumb happens to work for, and now you've got a mistake that is your liability. Not AI's liability. Yours.

The defense is the same cross-checking you'd apply to any unverified claim: trace it back to a primary source. If AI gives you a rule of thumb, ask where it comes from. Look it up in ASHRAE, in manufacturer data, in code tables. If AI says "values for X are between 1 and 4.3" and the ASHRAE handbook says the range is 0.8 to 6.1, you've caught a trap door. If AI can't tell you where its rule originated or what conditions define its boundaries, it's not a rule — it's a guess wearing a lab coat.

And here's the thing: if the rule has a real basis, AI will be able to give you sources. A lot of times, when you press on these things, AI just admits it made them up. That moment — the moment you catch it, correct it, and keep working toward your solution with verified data — that's where your money is made. That's engineering. The trap door becomes the opportunity. "Claude, where did this number come from?" might be the most valuable sentence in your entire workflow.

THE SPEED PARADOX

AI lets you iterate faster than you've ever been able to before. You can spin up a CFD simulation from a CAD model, watch the values as it runs, learn from the results, adjust parameters, and restart — in a fraction of the time it used to take. You can generate ten design options where you used to generate two. You can run sensitivity analyses that used to take weeks in a matter of hours.

This is genuinely transformative. And it creates a specific danger.

Speed amplifies whatever it's applied to. Fast iteration on a well-defined problem is engineering at its best. Fast iteration on a poorly defined problem is expensive chaos. If you skipped Chapter 3 — if you didn't define the problem properly, didn't challenge your constraints, didn't identify what you don't know — then AI speed becomes a centrifuge for bad decisions. You'll converge quickly and confidently on the wrong answer.

This is why the Einstein Method has to come before the tools. Define the problem first. Then apply speed to the definition. AI-assisted problem definition is spectacular — fast research, fast parametric studies, fast "what if" exploration. But if you let AI speed pull you past the definition phase into the solution phase before the problem is properly understood, you're accelerating in the wrong direction.

I'll be honest: this particular failure mode hasn't happened to me, because the Einstein Method prevents it. I define the problem before I touch any tool. By the time AI enters my workflow, the problem is constrained, the unknowns are identified, and the solution space is already narrowed. AI then operates within those boundaries at speed, which is exactly where it excels.

But I've watched it happen to others. Engineers who jump straight into AI-assisted design without spending the time on problem definition. They produce beautiful, detailed, completely wrong solutions in record time. The speed makes the mistake feel productive, which makes it harder to recognize as a mistake.

Don't let the speed seduce you. Define first. Then go fast.

I learned this lesson long before I was an engineer.

When I was twelve, my dad got me a boat. He could tell I loved to go fast and figured it was better for me to get that out of my system on the water before I turned sixteen and became a real danger on the road. It was a 1980 VIP Kona with a Johnson 140 on it. Very big motor for a very small boat.

I got my boater's safety course done, got my license in New York, and my dad started training me. How to dock. How to use reverse. The concept of not having brakes — which, if you've never driven a boat, is a more important detail than you'd think.

The day finally came when my dad decided I could take the family out and captain the ship. Everything was going fine. I pulled the boat out of the lift, took it out on the lake to open the throttle and get the engine warm before loading my elderly grandparents on board.

Coming back in to the dock, my dad starts in casually: "Too fast, Kyle. Too fast." I'm still about twenty feet out, so I swing the wheel and head back out to the lake to make another pass that'll please my father. He says, "Slow down, my son. You're making a lot of waves that are gonna make it harder."

I'm young and cocky. I swing it back in. A little slower, but still hot-dogging. My dad lost his patience. "SLOW DOWN. SLOW DOWN. You're gonna hit the dock."

I made the same mistake again. I thought: just go back out and take another pass so Dad doesn't yell at you. Still cocky. So I hit the throttle all the way down, planning to send it back out to the middle and bring it in clean. But for those of you who have driven a boat, you know — it steers from the back and pushes from the motor. I was in a half turn, full throttle, and the back end of the boat slammed into the end of the dock, sending the whole thing oscillating back and forth. My elderly grandmother nearly fell in the lake. Everyone was furious. I was an idiot.

I had tried to use speed to solve every problem in that boat. My dad's next words were: "The solution to every problem in a boat is not speed."

From that twelve-year-old to this engineer, and from me to you — the future of our discipline — speed is not the solution to every engineering problem you have. In this job, you have to be right. That means sometimes choosing to go slow to do the check right. Don't be rushed. Be right.

CROSS-CHECK ARCHITECTURE

Never use the same AI to check its own work.

This is the equivalent of peer review in traditional engineering, adapted for AI-assisted workflows. If Claude generated a calculation, don't ask Claude if the calculation is correct. Verify it with a different method — a hand calc, a different AI model, a deterministic software tool, a rule of thumb from your sniff test library. Or better yet, have AI run a deterministic tool for you. Then you only have to check the inputs, because you know the equations in the software are right. That's a much shorter verification path.

Build this into a habit:

Use AI to generate the first draft of a calculation or analysis. Then check the critical values against a completely independent method. If both methods agree, your confidence goes up — for this specific task, not globally. Agreement on one calculation doesn't mean you can relax on the next. If they disagree, don't average the results — investigate the discrepancy. One of them is wrong, and finding out which one teaches you something important about both methods.

Over time, build automated cross-checks into your workflows. Write scripts that validate AI output against known benchmarks. Create templates that include verification steps. The goal is a system where AI output never reaches a deliverable without passing through at least one independent check.

Think of it this way: you wouldn't have the author of a paper review their own paper. You'd send it to someone with a different perspective, different biases, different blind spots. AI tools have biases and blind spots too — they're just harder to see because the output is so polished. The cross-check is how you find them.

FROM ANNOYING LIE MACHINE TO POWERFUL TOOL

I want to tell you about the moment this shifted for me, because I think every engineer goes through some version of this transition and it helps to see what it looks like.

I had just started my engineering firm and was working with my second client. They had a drying process they wanted to automate and needed a sequence of operations for optimizing the drying cycle for energy efficiency. Easy, I thought. Map the psychrometrics, write up a control sequence, send them on their way.

Then they told me they needed me to actually write the program for their controller. One more step — I'd probably have to put it into some kind of standard controller platform, and I could get a controls engineer friend to help.

Then they told me the controller ran a very specialized version of C++ that was written for this specific old hardware. A language that essentially nobody knew.

This was early in the ChatGPT era. Maybe the 4o model. I was skeptical. AI had been mostly an annoyance up to that point — confident and wrong often enough that I didn't trust it for anything important. But I didn't have a lot of options. I couldn't find anyone who knew this obscure language, and the client needed the program.

So I described the PID loop I wanted to create. Very accurately — what the loop needed to do, what variables should be paired to what parameters, what the control logic should be for optimizing the drying sequence. I told the AI about the one-off language and asked what it suggested I do next.

It said: "I can write that for you."

So I let it try. And when I checked the code, the structure was exactly how I would have written it. Some syntax I didn't recognize — that was the obscure language — but there in the code were all the PID parameters, the variables I'd specified, paired to the right inputs, with the control logic structured the way I'd described it. The whole thing made sense to me.

That was the moment it went from annoying lie machine to powerful tool for engineers.

I had just created functional code in a language I didn't fully know, for a real client, on a real system, and it worked. But the key — and this is the point of telling you this story — is that I had the judgment to describe what I needed accurately, and I had the judgment to verify that what it produced was structurally correct. The AI handled the syntax of an obscure language. I handled the engineering.

If I hadn't known what a PID loop was supposed to do, I couldn't have described it. If I hadn't understood the control logic, I couldn't have verified the code. If I hadn't known what variables to pair to what parameters, I would have had no way to check whether the AI got them right.

The AI was the speed. I was the intelligence. And that combination — my judgment deployed at AI speed — produced something that neither of us could have produced alone.

That's the operating model for your career. Learn the engineering. Build the judgment. Then deploy it through AI at speeds your predecessors couldn't have imagined. The ones who try to skip the first two steps and jump straight to the speed will produce a lot of confident-looking work that doesn't survive contact with reality.

You're not going to make that mistake. You read Chapter 4.

CHAPTER 6: COMMUNICATION IS THE NEW PROGRAMMING

If you can't get your ideas out of your head and into someone else's — a client, a contractor, a code official, an AI — you will fail. And it will be your fault.

Not their fault for not understanding. Not AI's fault for misinterpreting your prompt. Not the contractor's fault for building it wrong. Yours. Communication is the responsibility of the communicator. Full stop. If the person on the other end didn't get it, you didn't deliver it. That's the standard, and it applies in every direction you'll ever communicate: to machines, to technical peers, and to people who have no idea what a heat exchanger does.

I'm putting this chapter here — after the Einstein Method, after calibration, after the operating philosophy of AI as speed — because communication is the medium through which every other skill in this book becomes useful. You can define a problem brilliantly. You can have flawless judgment. You can know exactly what needs to be built. None of it matters if you can't get it out of your head and into the world.

ANALOGIES TO LIVED LIVES

Here's the method. It's not a personality trait. It's not charisma. It's a repeatable technique that works every single time if you're willing to do the work.

When you need to explain an engineering concept to someone who isn't an engineer, your job is to find an analogy in their life — something they've already experienced, something they already understand — and connect the unfamiliar concept to that familiar experience. You're building a bridge from their world to yours.

Don't dumb it down. That's the mistake most engineers make. They think communicating with non-technical people means removing the substance, simplifying until nothing's left but a vague hand-wave. That's insulting to the listener and it doesn't work. People aren't stupid — they're unfamiliar. The gap between you and them isn't intelligence. It's vocabulary and context. Your job is to close that gap without removing the substance that matters.

Say you're explaining how a car's transmission works to a guy who rides a bicycle. You start where he already lives: "You know how your bike has gears, and when you shift up, you get more speed but you have to pedal harder? Same idea. But instead of a chain connecting two sets of gears, the gears in a transmission are touching each other directly..." And you walk him in. You don't say "it's complicated." You don't say "basically, it makes the car go." You give him the real mechanism, connected to something his hands and legs already know.

This works for everything. Heat transfer, fluid dynamics, pressure drop, control loops. There's always an analogy in someone's lived experience if you're willing to look for it. Your job is to find it. And the harder the concept, the more important it is that you find a good one — because a good analogy doesn't just help someone understand. It gives them a mental model they can use to think about the concept on their own after you've stopped talking.

Here's how I know the method works.

Over a ten-year relationship, a life partner of mine became practically a mechanical engineer in terms of understanding physical systems. Not because she studied engineering. Because I never held back the details when we talked about what I was working on. I'd explain the wind tunnel cooling system the way I'd explain it to anyone — find the connection to her experience, walk her in, give her the real physics.

After a few years of this, she started making connections I hadn't explicitly taught her. She'd see something in the real world and say, "I think this is the same as when you guys use the heat exchanger on the wind tunnels, right? The heat is flowing like this?" She was making second and third order connections — applying the concepts independently, in new contexts, without me prompting her.

That's how you know you're a good communicator. Not when someone nods along while you're talking. That's compliance, not comprehension. The test is: can they use what you've explained without you? Can they make new connections you didn't explicitly give them? If yes, you actually transferred understanding. If they just nod and then ask the same question differently next week, you gave them words, not knowledge.

THE SAME SKILL, DIFFERENT AUDIENCE

Here's why this chapter is called "Communication Is the New Programming."

When you prompt AI, you're doing the exact same thing. You're taking what's in your head — the problem, the constraints, the physics, the relationships that matter — and translating it into language that another intelligence can act on. The better your translation, the better the output. The more ambiguity you leave, the more the AI fills in the gaps with its own choices — sometimes good ones, sometimes terrible ones.

I designed an entire product line in 48 hours, and the reason the AI-assisted workflow produced accurate results was that I constrained it aggressively. I described the physics relationships. I specified how I wanted the work done and how I wanted it shown. I told it what standards to use. Anywhere I left ambiguity was a place where AI chose the route for me — and some of those choices would have been wrong if I hadn't been explicit.

That's not a prompting technique. That's communication. The same discipline that lets you explain a chiller plant to a building owner is the discipline that lets you get useful output from an AI model. Be specific. Be structured. Translate from your internal understanding into language the listener — whether human or machine — can act on.

The engineers who complain that AI gives them garbage are almost always the engineers who give AI garbage inputs. Vague prompts produce vague results. Ambiguous constraints produce unpredictable output. "Design me a cooling system" is not an engineering instruction any more than it would be if you said it to a junior engineer with no context. You'd never walk up to a first-year and say "design me a cooling system" and expect something useful. You'd define the load, the constraints, the performance targets, the budget, the operating conditions. AI needs the same clarity. It deserves the same quality of instruction.

Communication to AI, communication to peers, communication to non-technical humans — it's all the same fundamental skill. Understand what's in your head. Translate it into language the recipient can use. Verify that the translation landed.

MATT DOESN'T LISTEN

I want to tell you about the worst communicator I've ever worked with, because the lesson is the inverse of everything I just told you — and it cost real money.

Matt was a project manager for a contractor on a wind tunnel build. He'd done a few warehouse projects. Put up some storage sheds. Had some success in simple construction management. And he walked onto this project — a highly technical build where we're trying to make humans fly inside a machine — with the confidence of someone who thought warehouses and wind tunnels were basically the same thing.

His fundamental problem wasn't that he couldn't explain things. It was that he wouldn't listen.

He'd jump in on top of anyone talking and start speaking over them. In meetings, he'd never leave room for anyone to respond — just barrel through his list of what "I'm gonna do" and move on to the next item. When it came to understanding the cooling system, the controls sequence, the commissioning requirements — things that were genuinely complex and genuinely different from anything he'd done before — he treated it all like installing a refrigerator.

I started keeping a tally. Just a count, on a notepad, of how many times he talked over someone in a meeting. Not an estimate. An actual count. Over ten times in a sixty-minute meeting. Consistently.

Now, I want to be clear about something: I didn't trust my own perception alone. When you're frustrated with someone, it's easy to perceive them as worse than they are. Confirmation bias is real. So I checked with the tunnel owners — the people who were paying for this project, the mutual customer between my firm and Matt's company. I asked them, separately, what they thought of how things were going.

Full agreement. They found him frustrating. They weren't sure their project would succeed with him running the show. Three owners, independently saying the same thing: "That guy doesn't listen."

That gave me the cover I needed. This wasn't my perception. This was a documented, verified pattern that was threatening the project.

Several weeks later, during commissioning, in the job trailer. We had multiple teams mobilized. Hundreds of thousands of dollars in deployment costs for the testing phase. This was the window where everything had to come together. Matt starts listing what he's going to do — a list of things that would prevent the system from working in time to finish commissioning. He wanted to stop testing the machine to work on cosmetic building details. In the middle of a critical, time-limited, expensive commissioning deployment.

I started to explain, clearly and calmly, why this wouldn't work. The money at stake. The people already mobilized. The testing schedule. Matt started talking over me. Again.

"Matt, shut the fuck up."

He stopped.

"Here's some career advice for you. Everyone on this project is extremely pissed at how you talk over everyone. The ownership hates it. My team won't speak to you. No one wants to help you. This project is a disaster because you won't use your ears. You've talked over me and the owners over a hundred times just in the calls I've been on. It's logged. You are going to fail in this industry if you don't start listening. From now on, when I talk, you listen. When I finish, you can speak. I'll give you the same courtesy. This is the only way we get through this without you costing these people their life savings. They need this place open and making money. Now are you ready to get this back on track or what?"

The owners backed it up. They told him directly: we agree. You need to listen to Kyle. He knows this stuff and you don't. Follow his advice from here on out or we'll replace you.

Matt didn't take it well. But he changed his behavior. The project finished about two weeks later.

THE LESSON ISN'T THE CONFRONTATION

I'm not telling you that story so you can rehearse a dramatic speech for your next difficult coworker. The lesson is deeper than that.

First: listening is communication. Matt's problem wasn't his vocabulary or his technical knowledge. It was that he treated communication as broadcasting — one direction, outbound only. He was transmitting but never receiving. And because he never received, he was working with incomplete information on every decision he made. He didn't know what the cooling system needed because he never listened to the people who understood it. He didn't know what the owners were worried about because he talked over them every time they tried to tell him.

Every engineer who talks brilliantly but never listens is making the same mistake. You're working with half the data. You're missing the contractor's concern, the client's unstated priority, the tradesperson's field knowledge, the AI output that contradicts your assumption. Receiving is half the job.

Second: verify the pattern before you act on it. The tally wasn't petty. It was data. It confirmed the problem was real and quantifiable, not just an emotional reaction to a personality I didn't like. Checking with the owners verified it wasn't just one person's perception. When the confrontation happened, it was backed by evidence and supported by stakeholders. If I'd blown up at Matt based on a feeling, I would have been the asshole in the room. Because I'd documented and verified, I was the person who finally said what everyone was thinking — with the evidence to back it up.

That's communication about communication. Meta-level, but critical. And it's the same verification mindset from Chapter 1. Don't trust your perception. Check it. Especially when the stakes are high and emotions are involved.

Third: Matt's root problem wasn't his mouth. It was his humility. He'd built storage sheds and thought that qualified him for wind tunnels. He wouldn't listen because he didn't think he needed to. This is the same disease as the iFLY London engineer who deviated from a proven design against our recommendations and caused fifty thousand pounds in rework. Same disease as the engineer who ignores the principle that everything in a built system is there for a reason. They all share a common root: the assumption that they already know enough, and therefore don't need to receive information from others.

Listening requires admitting you might not have the full picture. That's "I don't know" as a communication skill, not just an engineering skill.

HOW TO GET BETTER AT THIS

Communication is a skill, which means it responds to practice. Here's what practice looks like.

Explain your current project to someone who isn't an engineer. Not the sanitized, dumbed-down version. The real project, with the real challenges, translated into their world. Watch their face. When their eyes glaze, you've lost the analogy — find a better one. When they lean in and ask a follow-up question, you've got them. The follow-up questions tell you which parts of your explanation are working and which aren't.

Do this regularly. Dinner conversations. Friends who ask what you do. Family members who've never understood your job. Each one is a practice session. And over time, you'll notice something: the process of translating engineering for non-engineers actually sharpens your own understanding. If you can't explain a concept simply, using analogies from everyday life, there's usually a part of it you don't fully understand yourself. The explanation exposes the gaps in your own thinking.

For AI communication, the practice is the same: be specific, be structured, and verify the output. When AI gives you something wrong, look at your prompt first. Was it ambiguous? Did you leave room for interpretation where you should have been explicit? Did you define the constraints, or did you assume the AI would know what you meant? Nine times out of ten, a bad AI result traces back to a communication gap in the prompt. Fix the prompt, fix the output.

For technical peers, the practice is documentation. Write your assumptions down. Write your constraints down. Write your reasoning down — not just your conclusions. The engineer who can articulate why they made a decision, not just what the decision was, is the engineer who gets trusted with bigger decisions. Because when something goes wrong — and it will — the reasoning is what lets the team learn from the failure instead of just reacting to it.

THE MULTIPLIER

Communication is the multiplier for everything else in this book.

The Einstein Method is a communication process — you're defining the problem in language clear enough that the solution becomes obvious. Calibrating your instruments is about building the vocabulary of judgment that lets you communicate what "wrong" looks like. Managing AI is managing communication — prompts, constraints, context, verification. Every skill in this book becomes more powerful in proportion to your ability to get it out of your head and into someone else's.

The engineers who will lead in the AI age aren't the ones with the best technical skills, though you need those. They aren't the ones with the most AI fluency, though you need that too. They're the ones who can translate between worlds — between the AI's output and the client's needs, between the code official's requirements and the contractor's reality, between the physics of the system and the budget of the owner.

You're not just an engineer. You're an interpreter. Between the technical and the non-technical. Between the digital and the physical. Between what AI produces and what actually needs to happen in the real world.

Get good at translating. It's the skill that makes all the other skills count.

CHAPTER 7: EATING THE LAYER CAKE

The world favors those who act.

Not those who study comprehensively and then act. Not those who complete the prerequisite sequence and then begin. Not those who wait until they feel ready. The world favors the people who start before they're prepared and figure it out on the way.

This chapter is about how you build your technical stack — coding, AI fluency, scripting, simulation, agent building — and the answer is going to frustrate anyone who likes tidy learning paths: you don't build it one layer at a time. You eat the whole cake sideways.

Think about how you actually eat a layer cake. You don't scrape off the frosting, eat it, then eat the top layer, then the middle layer, then the bottom. You cut a slice and take a bite that goes through every layer at once. The chocolate, the filling, the frosting, the base — all in one mouthful. That's how you learn a technical stack. Not coding first, then AI, then scripting, then simulation, then agents. You take a real problem and bite through all of them simultaneously.

THE OPENFOAM SATURDAY

Here's what that looks like in practice.

I needed CFD runs done — computational fluid dynamics simulations to verify product features for a potential manufacturing project. CFD costs about fifteen hundred to two thousand dollars per run if you hire it out. I needed a lot of runs. The math was ugly.

I reached out to some former colleagues from iFLY who were CFD experts. They walked me through the options: higher cost with quicker turnaround using SolidWorks CFD, or lower cost with a steeper learning curve using OpenFOAM, which is free, open-source, and runs on Linux. I looked into borrowing a SolidWorks seat from a former employer, but that conversation slowed down for various reasons and left me without a path forward.

It was a Saturday. I'd been thinking about this problem all week. I looked at my computer, looked at the OpenFOAM documentation, and thought: "I'm gonna do some nerd shit and see if I can get this free CFD software running. I've done ANSYS before. I'm smart. I know 85% of what I need to know. Let's just send it."

I loaded OpenFOAM on my Linux machine. First thing I discovered: OpenFOAM has no graphical interface. It's command line only. That's why my colleague had warned me about the learning curve. For a human sitting at a keyboard, navigating a complex simulation tool through nothing but typed commands is intimidating. There are no buttons to click, no menus to explore, no visual cues to guide you. Just a blinking cursor.

I fired it up anyway. Error. Version mismatch during the install.

"Claude — can you fix this OpenFOAM issue I'm having?"

Fixed. Package mismatch resolved. Software running. "What's next?"

And then the eureka hit me.

This is CLI software. Claude operates natively in CLI. The thing that made OpenFOAM intimidating to humans — no GUI, command line only, text-based configuration — is exactly what makes it perfect for AI. AI doesn't need buttons and menus. It processes text. It generates text. The command line is text. The software everyone avoided because it was hard for humans became the most AI-accessible option in the entire category.

"Claude, can you run OpenFOAM for me if I help define the problem? How would that work?"

It laid out the whole plan. Feed it the CAD files. Mark up drawings showing boundary conditions — just annotated images, the way engineers have communicated design intent for a century. Provide the problem conditions. Claude would handle the OpenFOAM configuration, the meshing, the solver setup, and the execution.

My career changed again in that moment.

I defined the problems. I set the boundary conditions based on my engineering judgment. I told Claude what physics to model. It configured the solver, generated the mesh, ran the simulation. When results came back, I evaluated them against my understanding of how fluid dynamics should behave in those geometries. If something looked off — a flow pattern that didn't match physical intuition, a pressure distribution that didn't make sense — I'd adjust the boundary conditions and run it again.

By about run twenty or twenty-five, I realized something: I was a CFD guy now. Not because I'd taken a CFD course. Not because I'd spent years mastering OpenFOAM syntax. Because I had the engineering judgment to define problems, evaluate results, and iterate — and AI handled the translation between my judgment and the software's requirements.

I did forty runs. For free. The cost was one computer and some setup time. The equivalent cost of hiring that work out would have been sixty to eighty thousand dollars.

I can now do hundred-thousand-dollar studies for a couple hours of my time. That's not a projection. That's my current reality.

WHAT JUST HAPPENED

Look at what that Saturday touched, all at once.

Coding and CLI literacy — I was working in a Linux terminal, troubleshooting package installations, navigating configuration files. LLM fluency — I was describing boundary conditions to Claude, iterating on solver setup, communicating engineering intent in natural language and having it translated into software configuration. Engineering domain knowledge — I was defining the physics, selecting boundary conditions, interpreting results against physical intuition built over twenty years. Tool building — I created a repeatable CFD workflow that I've used on every project since.

Nobody told me to learn these in order. Nobody said "first complete a Linux course, then learn OpenFOAM syntax, then practice prompt engineering, then try a simulation." A real problem forced all the layers at once, and I bit through them all in a single Saturday because I didn't have the luxury of doing it any other way.

That's the layer cake. Every real engineering problem cuts through multiple skill layers simultaneously. When you try to learn the layers separately — coding in isolation, then AI in isolation, then your domain tools in isolation — you're studying ingredients instead of cooking. The learning doesn't connect until you have a problem that forces the connections.

CLI-FIRST

There's a principle underneath the OpenFOAM story that I want to make explicit, because it will shape how you set up your entire technical workflow.

AI is natively not GUI. It's natively CLI.

Think about it. AI processes text. It generates text. It operates through language. The command line is a text interface. When you give AI access to a command line, you're giving it the interface it was built for. When you ask AI to work through a graphical interface — clicking buttons, navigating menus, interpreting visual layouts — you're asking a fish to run. It can sort of do it, but you're fighting the medium.

This is why I work CLI-first. Claude Code, Codex, Cowork — these tools give AI access to the filesystem. That means AI can read your project files, write code, run scripts, launch tools, modify configurations. It's the difference between a consultant who gives you advice over the phone and a consultant who shows up and does the work. Chat-based AI can talk about your project. CLI-based AI can touch your project.

And here's the progression that matters for your career: if you can tell an AI to do something from the command line, you can script that command. If you can script it, you can schedule it. If you can schedule it, you have an agent. The path from "I typed a command" to "I have thirty autonomous agents monitoring my projects" is a straight line through the CLI. Graphical tools don't offer that path.

This is why I push CLI literacy early. Not because I want to gatekeep with obscure terminal commands. Because the command line is the on-ramp to everything that comes next — and everything that comes next is agents, automation, and the Team of One that I'll describe in Chapter 10.

THE LAYERS, EATEN SIMULTANEOUSLY

Let me name the layers so you know what you're biting through. But remember: you don't learn these in sequence. You learn them by doing things that require all of them at once.

Coding literacy. Not mastery. Literacy. You need to understand the structure of code — what a function does, what a loop does, how data flows through a script — well enough to read what AI produces and evaluate whether the structure matches your intent. You don't need to write Python from scratch. You need to know enough Python to say "that loop is iterating over the wrong variable" when you read AI-generated code. Chapter 5 covered this: structural understanding across languages, not fluency in any single one.

LLM fluency. How to talk to AI so it produces useful output. This is Chapter 6 applied to machines — be specific, constrain aggressively, verify the output. But it's also pattern recognition: knowing which tasks AI handles well (boilerplate code, format conversion, documentation, configuration files) versus which tasks require heavy human judgment (boundary condition selection, design intent, failure mode analysis). The more you use AI, the sharper this sense gets.

Scripting and automation. Python, Bash, whatever your environment calls for. This is the bridge between one-off tasks and repeatable workflows. When you ask AI to do something once, that's useful. When you capture that workflow in a script that you can run again — or that an agent can run autonomously — that's a multiplier. Every script you write is a team member you're recruiting.

CAD and simulation. The domain-specific tools of your discipline. For mechanical engineers, that's SolidWorks, Revit, OpenFOAM, EnergyPlus, whatever your shop uses. AI is integrating into these tools rapidly. The engineers who understand both the tool and the AI layer will get dramatically more out of both. The OpenFOAM story is the proof: AI didn't replace the simulation software. It made the simulation software accessible in ways that changed the economics of using it.

Agent building. The frontier. This is where you take everything above — CLI fluency, scripting, AI orchestration — and build autonomous workflows that operate under your judgment but without your constant input. Supply chain monitoring agents. Code compliance checkers. Automated calculation verification. We'll go deeper on this in Chapters 8 and 10, but know that every layer below this one feeds into it.

THE EXTERNAL BRAIN

One more tool that ties the layer cake together: your decision log.

I use Obsidian — a note-taking system — connected to my CLI tools. Every project, every decision, every lesson learned gets logged. Why did we choose that chiller? What were the tradeoffs? What happened when we tried the alternative? What did the client actually care about versus what we assumed they cared about?

This sounds like bookkeeping. It's not. It's building an external brain that compounds over your entire career.

Five years of decision logs gives you — and your AI tools — an institutional memory that most firms lose when people change jobs. When a new project comes in with similar constraints to something you did three years ago, you don't have to rely on vague memory. You have the reasoning, the tradeoffs, the outcomes. And because it's all text-based and connected to your AI tools, you can search it, analyze it, and build on it. Your past decisions literally inform your AI-assisted future ones.

Start this habit now. Not when you have a perfect system. Not when you've evaluated every note-taking app. Pick one that works, start logging, and let it compound. The crappiest decision log you start today is infinitely more valuable than the perfect one you plan to start next year.

THE ON-RAMP

If you're sixteen and reading this, here's what you do this weekend.

Not next month. Not after you finish some prerequisite. This weekend.

Saturday morning: get a Linux terminal running. If you're on Windows, install WSL — Windows Subsystem for Linux. If you're on Mac, open Terminal. Get comfortable navigating files from the command line. Move around. Create directories. Break something. Fix it with AI's help. That process — break, diagnose, fix with AI assistance — is the fundamental loop you'll use for the rest of your career. Start building the muscle memory now.

Saturday afternoon: pick something real that annoys you. A spreadsheet that needs cleaning every week. A set of photos that need organizing. A calculation you do repeatedly for homework. Something small, something concrete, something you actually care about. Ask AI to build a tool that does it. Not a tool that explains it. A tool that does it. Get the AI to write a script, run it, and produce a result. If it doesn't work the first time — and it won't — iterate. That iteration is the learning.

Sunday: start a notebook. Obsidian, because it works with markdown files which are the basis of AI workflow. Log what you built yesterday. What worked. What didn't. What you'd do differently. This is entry number one in your decision log. It looks tiny and pointless right now. In five years (probably more like 5 months), it's a gold mine.

Next weekend, build another tool. Save it next to the first one. You now have two team members.

One tool per weekend for a year is fifty tools. Some will be garbage. Some will merge with each other. Some will become the foundation of something much bigger — my OpenFOAM workflow started as "I'm gonna do some nerd shit on a Saturday" and became a capability worth tens of thousands of dollars per project. You don't know which of your weekend experiments will turn into that. You don't need to know. The volume of attempts is what matters early. Quality comes from iteration. Iteration requires starting.

The distinction that matters: don't use AI to search for things. Use AI to do things. Searching is using AI as a better Google. Building is using AI as an engineering partner. The teenager who makes that jump — from asking AI questions to directing AI to build tools — has a compounding advantage over every peer who stays in search mode.

You are not just learning. You are building your team. Every tool you save, every script that works, every workflow you refine is a team member you're recruiting for the career you haven't started yet. By the time you graduate college, you should have a toolkit that makes you dramatically more productive than your classmates on day one. Not because you're smarter. Because you've been building your team since you were sixteen.

Start this weekend. The world favors those who act.

CHAPTER 8: MANAGING YOUR AI TEAM

You are the boss.

I don't mean that as a motivational platitude. I mean it as a job description. When you use AI in your engineering work, you are managing a team of very fast, very eager junior engineers who have read every textbook ever written but have never set foot on a jobsite. They'll work around the clock. They don't get tired. They produce polished deliverables at a speed that would have been physically impossible for a human team. And they will confidently give you work that is subtly wrong unless you scope it, checkpoint it, review it, and apply judgment at every decision point.

That's not a technology description. That's a management description. And if you've never managed anyone before — which is likely, given that this book is aimed at people early in their careers — then this chapter is going to teach you how. Because managing AI is managing. The principles are the same whether you're directing a human junior engineer or a machine. The only difference is speed.

WHAT YOUR BOSS IS ALREADY THINKING

I want to show you the view from the other side of the desk, because this will change how you understand your own value in the market.

I have engineers who work for me. Smart, capable people. And here's what keeps happening: by the time I finish defining a problem clearly enough to hand it off to one of them, AI has already given me two or three iterations. Not rough sketches — full working versions that I can evaluate, adjust, and refine. In the time it would have taken me to explain the context, the constraints, the deliverable format, and the quality expectations to a human engineer — and then wait days or weeks for them to produce a first draft that I'd need to revise anyway — AI has already delivered something I can act on.

That's not an exaggeration. That's my Tuesday afternoon.

So here's the question I need you to sit with: if AI can produce iterations faster than your boss can explain the problem to you, what is your value?

The answer isn't nothing. The answer is everything AI can't do.

AI can't go to the field. It can't stand in the mechanical room and notice that the piping doesn't match the drawing. It can't feel the vibration that tells you a bearing is going bad. It can't have the relationship with the contractor that makes them call you when they hit something unexpected instead of just covering it up and hoping nobody notices. It can't manage the building owner who's nervous about the budget. It can't sit across from a code official and negotiate an interpretation that saves the project six figures.

That's what I want from the engineers on my team now. I want them to understand the work — deeply, not superficially. I want them using AI themselves to produce results at speed. And I want them doing the things that require a human body, a human relationship, and developing human judgment. See things in the field. Manage people. Make the physical stuff happen. Don't get mired down in the numbers when AI can handle the numbers in a fraction of the time.

The junior engineer's job has fundamentally changed. It used to be: the senior gives you a problem, you calculate the answer, you bring it back in a week or two. That model is dead. If you're reading this book, your job description is something different: produce judgment, produce field intelligence, produce human coordination — and use AI to handle the computational work at the speed your boss expects.

Engineering no longer needs to take a long time to do hard and powerful things. The engineers who understand that — who operate at AI speed instead of fighting it — are the ones who thrive. The ones who insist on doing everything the slow way aren't paying dues. They're becoming bottlenecks.

THE BOSS MINDSET

Here's how the management model actually works in practice.

When I designed a complete product line in 48 hours — multiple models of rear door heat exchangers, each with proven component selections and mathematical performance proofs — I wasn't asking AI for advice. I was giving it instructions. The shift from "AI as advisor" to "AI as executor under direction" is the whole chapter in one sentence.

Tell it to do it, and record it over here.

That's the mindset. You're not having a conversation with AI. You're running a project. You're the project manager, the technical lead, and the quality control department. AI is the workforce.

The workflow looks like this. First, you define the scope — which is the Einstein Method from Chapter 3, applied to AI management. What's the problem? What are the constraints? What does the deliverable look like? What quality standards does it need to meet? You do this work before AI touches anything, the same way a good project manager writes a clear scope of work before the team starts.

Then you direct the execution. Not "figure this out for me" but "here are the physics relationships, here are the constraints, here's the format I want the output in, here are the standards to use." Remember Chapter 6 — anywhere you leave ambiguity is where AI chooses the route for you. Over-constrain. Let AI push back on constraints that are genuinely too tight. But you own the decisions. You approve the tradeoffs. You are the boss.

Then you review the output at checkpoints. Not at the end — along the way. If AI is generating a series of equipment selections, check the first one before it does the next twenty. If the first one is wrong, you've caught it early. If it's right, your confidence in the batch goes up. This is the same review cadence you'd use with a human team member, just compressed into minutes instead of weeks.

Then you verify through independent methods. This is the cross-check architecture from Chapter 5. AI-generated selection? Verify the critical values against a hand calc or a different tool. AI-produced calculation? Check the key numbers against your rules of thumb. If both methods agree, you're in good shape. If they disagree, investigate — don't average.

That's the management cycle: scope, direct, review, verify. It's not complicated. But it requires that you actually do it, every time, without getting lazy because the output looks professional.

THE BUSINESS REALITY

Here's a truth that most engineering programs never teach you and most young engineers don't think about until it's too late: you have to generate more income than you cost.

That's not cynicism. It's the fundamental economics of being a professional. Your firm pays you a salary. Your salary comes from project fees. Project fees come from billable work. If the revenue your work generates doesn't exceed what the firm spends on you — salary, benefits, overhead, training, desk space — you're a cost center, not an asset. And cost centers get cut when times get tight.

This has always been true. What's changed is that AI has radically altered the math.

A seventy-five-thousand-dollar-a-year engineer who can produce five hundred thousand dollars in billable work using AI workflows is a fundamentally different economic proposition than a seventy-five-thousand-dollar-a-year engineer who produces a hundred and fifty thousand in billable work doing things manually. The first engineer is a profit engine. The second is barely breaking even after overhead.

I'm not telling you this to scare you. I'm telling you this because when you understand the economics, the motivation for everything in this book changes. Every tool you learn, every workflow you build, every layer of the cake you bite through — it's not just professional development.

It's economic leverage. The gap between what you cost and what you produce can grow exponentially with AI tools, and that gap is what gives you negotiating power, career options, and eventually the ability to go out on your own if you want to.

Until you can see the direct line from your work to high-value revenue, don't get cocky — you're still learning. Humility about where you are in that curve is both accurate and respected. But keep your eyes on the line. The engineers who understand the business reality make better decisions about how they spend their time than the ones who think their job is just solving technical problems.

BE THE WATER

Let me address the fear directly, because I know it's there.

Some of you are reading this book excited. You see the tools, you see the speed, you see the career possibilities, and you want to get started. Good. Hold onto that.

Some of you are reading it scared. You hear "your boss is already comparing you to AI" and "some roles will go away" and your stomach drops. That's honest. I'm not going to tell you the fear is irrational, because it isn't entirely. Some roles will compress. The drafter who only drafts. The calculator who only calculates. The report-writer who only writes reports. Those functions are being absorbed by AI, and pretending otherwise would make this book dishonest.

But here's the reframe: be the water, not the rock.

The rock sits in the stream and tries to resist the current. The current doesn't care. Over time, the rock gets worn down and reshaped by the very thing it was trying to resist. That's the engineer who refuses to learn AI tools, who insists on doing everything the way it's always been done, who treats the transition as a threat to fight rather than a current to ride. The resistance isn't noble. It's erosion.

The water is different. The water is flexible and powerful at the same time. It shapes the landscape. It finds paths. It goes around obstacles and over them and through them. The engineer who adapts — who learns the tools, builds the workflows, develops the judgment to direct AI effectively — that engineer isn't being carried by the current. They're using it.

Here's something I want you to sit with: when I was a kid, AI chat was how I thought computers should just work. Talk to the machine and it does what you ask. We spent thirty years with keyboards and mice and menus and buttons, and the whole time the thing we actually wanted was to just tell the computer what to do and have it do it. Now we get to do that. This isn't a disruption — it's the fulfillment of a promise technology has been making for decades.

The fear goes away when you use the tools. Not when someone argues you out of it. Not when you read a reassuring article. When you sit down, use AI to solve a real problem, and realize it needs your judgment to work — the abstraction collapses and the reality takes over. The reality is that AI needs you. It needs your problem definition, your verification, your field knowledge, your human relationships. It's powerful and it's dependent on your direction. That should feel exciting, not threatening.

So yes, some roles change. Yes, the expectations have shifted. Yes, your boss is watching the numbers differently than they did five years ago. But the person who can define, direct, verify, communicate, and build relationships? That person doesn't compress. That person expands.

Make everything. The tools have never been this powerful. The leverage has never been this available. The people who lean into it — who become the water — will ride this to careers that previous generations couldn't have imagined.

The people who fight it will watch the stream flow around them while they wonder what happened.

Don't be the rock.

CHAPTER 9: THE RISK SPECTRUM

Not every problem needs 100% certainty. And knowing when to move at less than 100% is one of the most valuable judgment calls you'll ever develop.

There's a tension running through this entire book. Chapter 3 says define the problem thoroughly before you solve it. Chapter 5 says use deterministic tools for safety-critical calculations. Chapter 7 says act fast, the world favors those who act. These aren't contradictions — they're different points on the same spectrum, and your job is to know which point applies to the situation in front of you.

That's the risk spectrum. On one end: move fast, accept imperfect information, learn through iteration. On the other end: stop, verify everything, get it right before you act. Both ends are correct, depending on what you're doing. The skill is knowing which end you're on.

THE 85% RULE

A lot of decisions in engineering don't need certainty. They need enough.

When I sat down on that Saturday and decided to install OpenFOAM — free CFD software I'd never used, on a Linux system, with nothing but AI to help me figure it out — I knew about 85% of what I needed to know. I understood fluid dynamics. I'd done ANSYS simulations before. I knew what CFD was supposed to produce and how to evaluate whether results were physically reasonable. What I didn't know was the specific software, the CLI workflow, the configuration syntax.

"I know 85% of what I need to know. Let's just send it."

That was the right call. The consequence of being wrong was a wasted Saturday. The potential upside was a career-changing capability. The missing 15% was tool-specific, not domain-specific — exactly the kind of gap that AI fills well. The risk-reward calculation was obvious.

That's the 85% rule: when the consequences of being wrong are low and recoverable, and you have enough domain knowledge to evaluate the results, move forward with imperfect information. Don't wait until you've studied every tutorial and read every manual. Act, hit the wall, learn from the wall, adjust. The world favors those who act, and the engineers who wait for 100% certainty before every decision move so slowly that opportunities pass them by.

But here's the critical caveat — and this is where a lot of people get the 85% rule wrong. The 85% has to be the right 85%. Knowing 85% of the software syntax but 0% of the underlying physics is not the same as knowing 85% of the physics but 0% of the software syntax. The first situation is dangerous — you can't evaluate the results, so you can't catch errors. The second situation is manageable — you understand what the results should look like, and AI can bridge the tool-specific gap. The domain knowledge is the 85% that matters. The tool knowledge is the 15% that AI handles.

THE OTHER END: ARE WE RUN READY?

Now the other end of the spectrum.

When I was commissioning vertical wind tunnels, before every test run, I'd ask every person on the team the same question: "Are we run ready?"

Not a rhetorical question. Not a formality. Every person was asked individually, and every person was expected to give a real answer. Because at that point, we were about to do something with real consequences — start a machine that generates enormous forces and moves serious quantities of air, with humans inside it. The consequences of getting it wrong weren't a wasted Saturday. They were equipment damage, project delay, and potential safety incidents. In a vertical wind tunnel, a loose screw and washer on the ground become a bullet with a parachute.

The answers I got from that thirty-second ritual prevented hundreds of mistakes. "There's debris in the flow path." "Chiller system isn't ready — you have limited heat rejection capacity." "Tony needs five more minutes, he's mid-update on the parameters." No single person held all of that information. The only way to surface it was to ask everyone.

That's the other end of the spectrum. When the consequences are high and the action is irreversible or expensive to undo, you don't move at 85%. You verify. You check. You ask every person who might know something you don't. You pause.

The number of times I've paused before a startup and thought "actually, I should probably close that valve first" or "let me restrict that parameter before we go" is in the hundreds. Each of those pauses cost me ten seconds. Some of them saved hours of rework. Some of them may have prevented something worse.

"I CAN'T MAKE IT BROKENER"

And then there's the moment in Qatar. The bent pin inside the VFD card slot. The team staring at weeks of delay. The mood heavy in the room.

"If it's already broken, I can't make it brokener."

I pulled the card out and went to work with a pen clip. Twenty-five minutes of improvisation in hundred-degree heat, performing what amounted to microsurgery on industrial electronics. If it didn't work, we were exactly where we started — stuck, waiting for a replacement part that would take weeks to arrive in the Qatari desert. If it did work, we were back in business.

That was a 100% appropriate application of the 85% rule. The downside was zero — the component was already nonfunctional. The upside was enormous — saving the entire commissioning schedule. The judgment call was: what's the worst that happens? I've already absorbed the worst case. Anything I try from here is pure upside.

Both of these approaches — "are we run ready?" and "I can't make it brokener" — are correct applications of risk judgment. They look like opposite philosophies, but they're the same skill applied to different consequence profiles. The pen clip moment required the judgment to recognize that the situation was already at maximum downside. The run ready ritual required the judgment to recognize that the situation had significant downside remaining. Knowing which frame applies is the skill.

FAILING PROPERLY

You will fail. Not might. Will. Despite everything in this book — the sniff test, the cross-checks, the run ready ritual, the verification habits — something will go wrong. AI will produce something subtly wrong that you'll miss. You'll make a call with incomplete information and it won't play out the way you expected. You'll move wires on a live system without thinking about where that 24 volts is going to land and fry a PLC channel.

I already told you the PLC story in Chapter 1, but I want to come back to it here because it illustrates something about risk and failure that the first telling didn't fully address.

When I fried that board, the failure was a direct result of operating without the pause. I moved wires on a live system — 24 volts DC, not dangerous to a person, but absolutely capable of killing an electronic channel if it lands where it's not expected. I didn't take the "do nothing" moment. I didn't pause. I acted at speed on a task that deserved a moment of verification.

The consequence was medium — money and time lost, nobody hurt, fixable with a replacement board. The response was proportional: own it immediately, get the fix in motion, share the lesson with the team so everyone else builds that pause into their process.

Here's the template. Memorize it. You'll use it more than once in your career.

Own what happened. Explain the reasoning behind the call you made — not to excuse it, but to give the team the context they need to learn from it. Don't blame anyone or anything, including AI, including the tool, including the circumstances. State the fix. State the process improvement that prevents a repeat. Execute.

Then move on.

Living in the "whodunit" can feel productive, but it isn't. Your team doesn't want to watch you grovel. They don't want a three-day post-mortem about who's at fault. They want to see you own it, fix it, and get back to work. That's what earns respect. That's what earns trust.

And I mean this in the most literal sense possible: you can earn credibility by failing properly. The engineer who breaks something, owns it immediately, shares the lesson, and adds a verification step to prevent it next time is an engineer people want on their team. The one who hides mistakes, deflects blame, or — God forbid — blames the AI tool... that person is a liability regardless of how technically brilliant they are.

This connects directly to Rule Number One. When AI output causes a failure that you missed, the failure template still applies. Own the fact that you approved the output. Identify why your verification process didn't catch it. Improve the process. "The AI told me to" is not in the template. It will never be in the template.

CALIBRATING YOUR RISK SENSE

How do you develop the judgment to know where you are on the spectrum? How do you learn to tell the difference between a "send it" situation and a "pause and verify" situation?

The same way you calibrate everything else in this book: exposure, repetition, and deliberate practice.

Start by classifying decisions before you make them. Not formally — you don't need a risk matrix and a color-coded spreadsheet. Just ask yourself: what happens if I'm wrong? If the answer is "I waste some time and try again," you're on the 85% end. If the answer is "equipment gets damaged" or "the client loses money" or "the system doesn't meet code," you're on the verification end. If the answer is "someone could get hurt," you're at the absolute verification end — you don't move until you're certain, and even then you ask everyone around you if they see something you don't.

Over time, this classification becomes automatic. You develop a sense for consequence severity that runs in the background of every decision. It's the same pattern recognition you're building with rules of thumb and sniff tests — just applied to the decision about how carefully to approach a problem rather than the problem itself.

AI actually helps you practice this. When you're working with AI on a low-stakes task — generating a first draft of a report, exploring design options, running sensitivity analyses — practice moving fast. Build the habit of iteration speed. When you shift to high-stakes work — final calculations, equipment selections that will be purchased, designs that will be built — practice shifting into verification mode. Build both speeds. The engineer who can only move fast makes dangerous mistakes. The engineer who can only move carefully misses opportunities. You need both gears.

THE SPECTRUM IN ONE SENTENCE

Move fast on iteration. Pause on commitment.

That's it. When you're exploring, defining, testing, prototyping, learning — speed is your friend. When you're about to submit, purchase, build, stamp, or start a machine — the pause is your friend. Everything else in this chapter is just calibration on where the line between those two modes falls for any given situation.

And if the thing is already broken? You can't make it brokener. Get the pen clip.

CHAPTER 10: THE TEAM OF ONE

Here's where this is all going.

You will be a team of one or two people, running thirty to a hundred AI agents. Most of them will be autonomous. They'll be working in the background while you're in the field, while you're in a meeting, while you're sleeping. And they'll be bringing you information constantly:

"This tool is done."

"Your report is ready."

"The material arrived at the shop. It's being loaded into processing, just as expected."

"WARNING: Supply chain disruption detected for component TY6R. Suggesting Supplier B for the next few months while the disruption clears. Expect warranty increase of 0.25%. Approve, or shall I suggest alternatives?"

That last one is a real business decision, surfaced by an autonomous agent that has enough context to identify the disruption, evaluate alternative suppliers, quantify the tradeoff, and present you with a specific recommendation. Your job is to make the call. Approve, reject, or ask for more options. The agent does the research. You apply the judgment.

Your boss will be the CEO. Your company will be ten to twenty people and do hundreds of millions in business. The engineering firms that still need a hundred and fifty people to design twenty-five buildings a year will not be able to compete with firms that achieve the same output with a fraction of the headcount and dramatically lower overhead.

This isn't a prediction about some distant future. I'm already doing the work of an entire engineering team from a previous employer. Supply chain monitoring, project management, design iteration, report generation — things that used to require multiple full-time positions are running through workflows I've built with AI tools. The tools get better every month. The agent capabilities expand every quarter. The trajectory is a straight line from where we are now to the vision I just described.

THE ARCHITECTURE IS THE MOAT

When someone hires you in the AI age, they're not hiring just you. They're hiring your whole team.

Your agents. Your workflows. Your toolchain. Your judgment applied at scale. Two engineers with identical degrees, identical experience, identical technical knowledge — but one has built a team of fifty agents that handle supply chain monitoring, report generation, design iteration, code compliance checking, and project management — those two engineers are not competing for the same jobs. They're in different economic categories entirely.

The architecture is the moat. Just like a software company's value lives in its codebase and infrastructure, the AI-era engineer's value lives in the agentic architecture they've built — the agents they've configured, the workflows they've refined, the integrations they've connected, the judgment they've encoded into automated verification checks. This is compound interest applied to your career. Every agent you build, every workflow you automate, every verification step you encode adds to an architecture that makes you more capable and more valuable over time.

This is why I told you in Chapter 7 to start building your tools in high school. Every year of architecture development compounds. The person who starts building at seventeen has a five-year head start on the person who starts at twenty-two. And unlike traditional career advantages — which often depend on who you know or where you went to school — this one is entirely within your control. You build it with your own hands, one tool at a time, one agent at a time, one weekend at a time.

THE JUDGMENT PREMIUM

Here's what doesn't change in this future: the human is still at the center.

The agents handle execution. They monitor, analyze, generate, report. They do the computational work at speed and scale that no human team can match. But every output still flows through a judgment layer, and that layer is you. The supply chain agent suggests Supplier B, but you decide whether the warranty tradeoff is acceptable for this specific project. The design agent generates ten options, but you evaluate which ones are physically constructable and which ones look good on screen but can't actually be built. The compliance agent checks the code, but you know from field experience which code interpretations the local official actually enforces and which ones they're flexible on.

That judgment premium doesn't just persist in the AI age — it increases. When a hundred agents are producing output constantly, the quality of your judgment in triaging that stream is the bottleneck for the entire operation. Your sniff test, your eye-crometer, your rules of thumb, your field experience, your understanding of design intent — all of the calibration you've been building since Chapter 4 — that's what makes the difference between an architecture that produces excellent work and one that produces polished garbage at scale.

The agents make the quantity possible. Your judgment makes the quality possible. Together, they produce something that neither could achieve alone — high-quality engineering output at a volume and speed that reshapes the economics of the profession.

WHAT THIS LOOKS LIKE, TODAY

Let me make this concrete with what's already happening.

The RDHx product line I described in Chapter 8 — an entire lineup of rear door heat exchangers, designed in 48 hours with proven component selections and mathematical performance proofs — that was one person directing AI tools across multiple engineering domains simultaneously. Competitive landscape research, coil selections through manufacturer portals, custom fan selection software built and used within the same workflow, AMCA standard compliance, assembly specifications. Every function that would have required a multi-person team, compressed into a two-day sprint by one engineer with the judgment to direct the work and verify the results.

The OpenFOAM workflow from Chapter 7 — forty CFD runs worth sixty to eighty thousand dollars in equivalent consulting fees — that was one person and a CLI-based AI tool. The capability didn't exist a year before it happened.

Now it's a permanent part of my toolkit, available for every project.

These aren't isolated moments. They're data points on a curve. Each one adds a capability to the architecture. Each capability enables the next one. The product line design workflow feeds into the CFD workflow feeds into the supply chain workflow feeds into the project management workflow. The architecture grows, and as it grows, the ratio between what one person can produce and what one person used to be able to produce stretches further and further.

THE STRUCTURE OF THE FUTURE FIRM

The engineering firm of the near future doesn't look like what you've seen. It's not a big office with rows of desks and a hierarchy of junior engineers feeding work up to associates who feed it up to partners. That model was designed for a world where human throughput was the constraint — you needed bodies to do the work, so you hired bodies and built management structures to coordinate them.

When AI removes the throughput constraint, the structure inverts. You need fewer bodies and more judgment. The value concentrates in the people who can define problems, direct agents, verify results, maintain client relationships, and make the field calls that no agent can make. Ten to twenty of those people, each running their own agentic architecture, each bringing their own judgment and relationship network, can do the work that used to require a hundred and fifty.

The economics are brutal for the old model. If one firm can deliver equivalent quality with one-tenth the head-count — and therefore dramatically lower overhead — the pricing pressure will compress margins for everyone who hasn't made the transition. This isn't a theoretical possibility. It's the same disruption curve that's played out in every industry that AI has touched, and engineering isn't immune.

I'm not saying this to scare you. I'm saying it because I want you on the right side of it. The engineers who build the architecture will be the ones setting the terms. The ones who don't will be the ones responding to them.

BUILDING YOUR ARCHITECTURE

You've been building it since Chapter 7.

Every tool you saved from your weekend builds. Every script that automates a calculation you used to do by hand. Every workflow that connects two systems that used to require manual transfer. Every verification check you've encoded so it runs automatically instead of requiring your attention. Every decision log entry that gives your future agents context for making better recommendations.

That's your architecture. It started small — a Python script here, a CLI workflow there, a saved prompt template that produces better output than starting from scratch. It grows as your career grows. As you encounter new problem types, you build new agents to handle them. As you learn from failures, you encode the lessons as automated checks. As you develop judgment in new domains, you extend the architecture to operate in those domains.

By the time you're five years into your career, your architecture should be handling the routine work that consumes most junior engineers' time — the boilerplate calculations, the standard report sections, the equipment schedule formatting, the code compliance checklists. That frees you up for the high-judgment work: the field visits, the client conversations, the design decisions that require context no agent has, the problem definitions that require the Einstein Method applied by a human who understands the physics and the politics and the budget.

By ten years, you're running agents across entire project lifecycles. Design agents, procurement agents, construction administration agents, commissioning verification agents — all operating under your judgment, all producing work that bears your professional standard.

By twenty years, you are the firm. Or you're one of ten or twenty people who are the firm, each of you bringing a judgment domain and an agent architecture that, combined, can deliver what used to require an organization of hundreds.

THE CLOSING

I want to end this book where I started it — with the Lyon quote that's been running through every chapter.

"If engineering school is supposed to teach you anything at all, it's that you know nothing. But it teaches you how to verify."

That was true when a project manager in France said it to me during my first wind tunnel commissioning. It's true now. And it will be true when you're running thirty agents and making decisions at a pace your predecessors couldn't have imagined.

You know nothing — not with certainty, not without checking. The AI doesn't know anything either — it predicts, it generates, it produces work product at extraordinary speed, but it doesn't know. The knowing comes from verification. From the sniff test, the cross-check, the field visit, the pause before the button, the mentor's voice in your head saying "why did you use that value and not this one?"

That's your job. That's your value. That's what makes you the human at the center of a future that needs humans at the center.

Build your judgment. Build your tools. Build your team.

You know nothing. But you know how to verify. Now go build.

APPENDIX A: THE WEEKEND BUILD GUIDE

Chapter 7 gave you the philosophy. This appendix gives you the first four projects. Each one is bigger than the last. Each one builds on what came before. Do them in order.

• • •

PROJECT 1: SET UP YOUR NOTEBOOK

Before you build anything, build the system that remembers what you build.

Pick a weekend. Set up Obsidian — or whatever markdown-based note system works for you — on your computer. Create a vault. Make a folder for your first project. That's it. Don't plan a folder structure. Don't design a system. Don't spend three hours picking the perfect template. Just start a notebook and put something in it.

From this point on, everything you learn, every tool you build, every project you attempt gets logged here. Not polished. Not pretty. Just captured. Let AI help you keep track of decisions, notes, and lessons as you go — that's what it's good at. The goal is a searchable, linked record of your engineering thinking that compounds over your entire career. AI tools can read this. Future-you can search this. The structure will evolve on its own as you use it — you'll build your own categories, your own connections, your own markdown links between ideas. It starts ugly and small. That's fine. Start it.

When this project is done, you have: a working notebook, your first project folder, and the habit of capturing before you move on.

• • •

PROJECT 2: RUN A RESEARCH PROJECT

Pick something you're curious about in your discipline — or something you need to learn for school or work. A material property you don't understand. A system type you've heard of but never studied. A code requirement that confuses you. Something with enough depth that you can't answer it in a single search.

Use AI to research it and build you a reference document. Have it pull together manufacturer data, code references, textbook explanations, and real-world application examples. Your job is to direct the research, evaluate what comes back, and push back where it's wrong or incomplete. "That's not specific enough." "Where did that number come from?" "Add the edge cases where this breaks down." You're the editor. AI is the writer. Complain, correct, redirect until the document actually teaches what you need to know.

Save it in your notebook. You now have a reference manual on that topic that you built together — not by writing it from scratch, but by directing AI and applying your judgment to shape the result. That's the workflow for the rest of your career. Get used to it here.

When this project is done, you have: a reference document you built with AI, practice directing and correcting AI research output, and your first experience evaluating AI claims against primary sources. You've also practiced the reading speed techniques from Chapter 4 — you had to move through a lot of material with purpose.

• • •

PROJECT 3: BUILD A SOFTWARE TOOL

Pick a real problem. Something repetitive in your schoolwork or job. A unit conversion you do constantly. A calculation you run by hand every week. A data format you have to clean up before you can use it. Something small enough to finish in a weekend, real enough that you'll actually use it again.

Direct AI to build a script that solves it. Not a script that explains the solution — a script that does the work and produces a result. Run it. It won't work the first time. Debug it with AI's help. Iterate until it does what you need. Save it somewhere permanent with a clear name and a note in your notebook about what it does, when you'd use it, and what you learned building it.

You don't need to understand every line of the code. You need to understand the structure well enough to know whether it's doing what you asked. If it produces output, check the output against a known answer. If it processes data, spot-check the results. You're the boss. The script is the worker. Verify its work.

When this project is done, you have: a working tool that saves you time on a real task, your first experience directing AI to build something functional, and your second team member sitting in your toolkit waiting to be used again.

• • •

PROJECT 4: BUILD A HARDWARE TOOL

This one leaves the screen.

Pick a physical problem. A sensor that monitors something in your environment — temperature, humidity, vibration, light levels, power consumption. A controller that automates something manual. A measurement rig for a project. Something that interacts with the real world and produces data you can use.

Use AI to help you design it, spec the components, write the firmware or control logic, and troubleshoot the build. The hardware might be an Arduino, a Raspberry Pi, a PLC trainer, or whatever platform makes sense for your discipline and budget. The point isn't the platform — it's the experience of directing AI to help you make something physical work.

This is where every layer of the cake gets bitten at once. You'll need to communicate clearly with AI about physical constraints that it can't see. You'll need to evaluate whether its electrical or mechanical suggestions make sense in the real world. You'll need to debug problems that exist in physical space, not just in code. You'll need to read datasheets and manuals at speed. You'll need your sniff test for whether a sensor spec or a wiring diagram looks right.

When this project is done, you have: something physical that works, experience bridging the gap between AI's digital world and engineering's physical one, and proof — to yourself and anyone watching — that you can make real things happen.

• • •

THE PATTERN

Notebook → Research → Software → Hardware.

Each one is harder. Each one is more real. Each one requires more judgment and more verification. And each one adds a capability to your architecture that you'll build on for years.

After these four, you don't need an appendix anymore. You'll see the next project yourself — because the problem that needs solving will show up in your work, and you'll already know the pattern for attacking it. That's the whole point. These four aren't the destination. They're the on-ramp.

Go build.

APPENDIX B: THE SNIFF TEST CHEAT SHEET

I'm not going to give you one.

Not because I don't have rules of thumb — I have hundreds of them, built over twenty years of mechanical engineering. But my rules of thumb are calibrated to my discipline, my experience, my climate zones, my equipment preferences, my project types. If I gave you a cheat sheet of HVAC rules of thumb, it would be useful to maybe ten percent of the people reading this book. The structural engineers would ignore it. The electrical engineers would ignore it. The civil engineers would ignore it. And the mechanical engineers who work in different applications than mine would need different numbers anyway.

Here's what I want you to do instead: go get yours from a person.

Find the old head in your office. The engineer in your social circle you respect. The mentor you found in Chapter 4. Sit down with them and ask: "What are the rules of thumb you carry in your head? The ones you use every day to check whether a number is in the right ballpark?"

Every experienced engineer has these. And every experienced engineer loves to talk about them. This is the stuff they're most proud of — the judgment they've built over decades, distilled into quick checks that let them evaluate work at a glance. Ask, and they will pour it out.

I used to work for a structural engineer who could look at a situation and spec a beam from memory. He'd put a finger in the wind and say, "Seven kips, twenty-foot span — that's a W16x26, probably. W21x44 is overkill. They planned for this. Proceed." And he'd be within a size or two of the final answer. Not precise enough to stamp a drawing, but more than enough for the back-of-napkin math we needed in the moment. That man had the AISC steel manual living in his head. He'd built that instrument over decades, one beam at a time, and it was beautiful to watch.

That's what you're building toward. And the fastest way to start is to learn from someone who's already there.

So here's your assignment. Find your version of that structural engineer. Buy them a coffee. Ask them for their cheat sheet — the rules of thumb, the quick checks, the numbers they carry in their head. And don't stop there. Ask them where each rule breaks down. What are the edge cases? What conditions make the rule unreliable? When has it led them astray? The rule of thumb is useful. Knowing its boundaries is what makes it safe. Write them down — the rules and the exceptions together. Memorize the ones that apply to your work. Start using them to check every number that crosses your desk.

That's your sniff test cheat sheet. It's different for every discipline, every application, every engineer. And it's worth more than anything I could print on a page.

ACKNOWLEDGMENTS

To my father, who was an engineer before I was anything. He raised me around the mindset that became this book — the curiosity, the drive to understand how things work, the belief that the answer is always findable if you're willing to chase it far enough. He put me on a boat at twelve and taught me that speed isn't the solution to every problem. I'm still learning that one.

To Mike, who gave me the bedrock. Two and a half years at JB&B, and every question I asked got a real answer. He didn't hand me numbers — he handed me reasoning. The way I think about engineering started in his office, tracing piping on drawings and asking why. Everything in this book that works traces back to the foundation he built.

ABOUT THE AUTHOR

Kyle Benesh is a multi-discipline mechanical engineer with twenty years of experience spanning building systems, vertical wind tunnels, computational fluid dynamics, and product design. He has worked at JB&B Engineering in New York City, spent nearly a decade at iFLY designing and commissioning the cooling systems for vertical wind tunnels on four continents, and now runs his own engineering firm and is the lead engineer for several manufacturing ventures. He builds things with AI every day and believes the future of engineering belongs to the people who pair human judgment with machine speed. He lives and works from his computer, his CLI, and occasionally a job trailer.

www.ingramcontent.com/pod-product-compliance
Lightning Source LLC
LaVergne TN
LVHW022108200926
848481LV00003B/1109